Complete Handbook
of Electric Motor Controls

John E. Traister

PRENTICE-HALL, Englewood Cliffs, New Jersey 07632

Library of Congress Cataloging-in-Publication Data

Traister, John E.
    Complete handbook of electric motor controls.

    Includes index.
    1. Electric controllers.   2. Electric motors—
Starting devices.   3. Electric motors—Electronic
control.   I. Title.
TK2851.T65   1986   621.46′2   85-28121
ISBN 0-13-160938-6

Editorial/production supervision: *Raeia Maes*
Cover design: *Photo Plus Art*
Manufacturing buyer: *Rhett Conklin*

© 1986 by Prentice-Hall
A Division of Simon & Schuster, Inc.
Englewood Cliffs, New Jersey 07632

Figures 1.1, 1.2, 1.4–1.11, 4.1–4.3, and 13.1–13.11 are from
John E. Traister, *Handbook of Electric Motors: Use and Repair,*
© 1984. Reprinted by permission of Prentice-Hall.

Printed in the United States of America

10   9   8   7   6   5   4   3   2

ISBN   0-13-160938-6   025

Prentice-Hall International, Inc., *London*
Prentice-Hall of Australia Pty. Limited, *Sydney*
Editora Prentice-Hall do Brasil, Ltda., *Rio de Janeiro*
Prentice-Hall Canada Inc., *Toronto*
Prentice-Hall Hispanoamericana, S.A., *Mexico*
Prentice-Hall of India Private Limited, *New Delhi*
Prentice-Hall of Japan, Inc., *Tokyo*
Prentice-Hall of Southeast Asia Pte. Ltd., *Singapore*
Whitehall Books Limited, *Wellington, New Zealand*

Contents

3   Electricity Basics   81

4   Electric Motors   98

5   Manual Full-Voltage Motor Starters   114

6   AC Magnetic Starters   124

7   AC Magnetic Reversing Controllers and Combination Starters   130

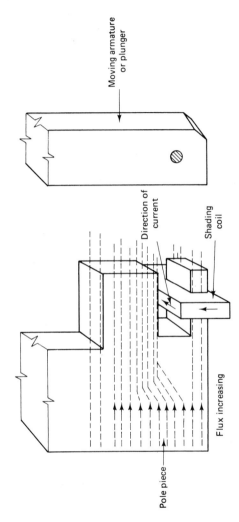

Figure 1-1 Section of pole face with current in clockwise direction.

Moving armature
or plunger

Direction of
current

Shading
coil

Flux increasing

Pole piece

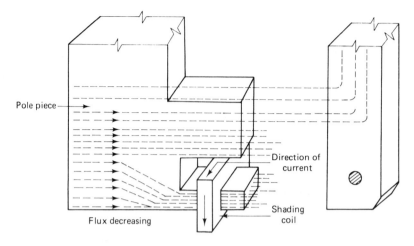

Figure 1-2   Section of pole face with current in counterclockwise direction.

The magnetomotive force produced by the coil opposes the direction of the flux of the main field. Therefore, the flux density in the shaded portion of the iron will be considerably less, and the flux density in the unshaded portion of the iron will be more than would be the case without the shading coil.

Figure 1-2 shows the pole with the flux still moving from left to right but decreasing in value. Now the current in the coil is in a counterclockwise direction. The magnetomotive force produced by the coil is in the same direction as the main unshaded portion but less than it would be without the shading coil. Consequently, if the electric circuit of a coil is opened, the current decreases rapidly to zero, but the flux decreases much more slowly due to the action of the shading coil.

Electrical ratings for ac magnetic contactors and starters are shown in Fig. 1-3.

OVERLOAD PROTECTION

Overload protection for an electric motor is necessary to prevent burnout and to ensure maximum operating life. Electric motors will, if permitted, operate at an output of more than rated capacity. Conditions of motor overload may be caused by an overload on driven machinery, by low line voltage, or by an open line in a polyphase system, which results in single-phase operation. Under any condition of overload, a motor draws excessive current that causes overheating. Since motor winding insulation deteriorates when subjected to overheating, there are established limits on motor operating temperatures. To protect a motor from overheating, overload relays are employed on a motor control to limit the amount of current drawn. This is *overload protection,* or *running protection.*

The ideal overload protection for a motor is an element with current-sensing properties very similar to the heating curve of the motor (Fig. 1-4) which would act to open the motor circuit when full-load current is exceeded. The operation of the protective device should be such that the motor is allowed to carry harmless overloads but is quickly removed from the line when an overload has persisted too long.

Fuses are not designed to provide overload protection. Their basic function is to protect against short circuits (overcurrent protection). Motors draw a high inrush current when starting and conventional single-element fuses have no way of distinguishing between this temporary and harmless inrush current and a damaging overload. Such fuses, chosen on the basis of motor full-load current, would "blow" every time the motor is started. On the other hand, if a fuse were chosen large enough to pass the starting or inrush current, it would not protect the motor against small, harmful overloads that might occur later. Dual-element or time-delay fuses can provide motor overload protection but suffer the disadvantage of being nonrenewable and must be replaced.

The *overload relay* is the heart of motor protection. It has inverse-trip-time characteristics, permitting it to hold in during the accelerating period (when inrush current is drawn), yet providing protection on small overloads above the full-load current when the motor is running. Unlike dual-element fuses, overload relays are renewable and can withstand repeated trip and reset cycles without need of replacement. Overload relays cannot, however, take the place of overcurrent protective equipment.

The overload relay consists of a current-sensing unit connected in the line to the motor, plus a mechanism, actuated by the sensing unit, which serves, directly or indirectly, to break the circuit. In a manual starter, an overload trips a mechanical latch, causing the starter contacts to open and disconnect the motor from the line. In magnetic starters, an overload opens a set of contacts within the overload relay itself. These contacts are wired in series with the starter coil in the control circuit of the magnetic starter. Breaking the coil circuit causes the starter contacts to open, disconnecting the motor from the line.

Overload relays can be classified as being either thermal or magnetic. *Magnetic overload relays* react only to current excesses and are not affected by temperature. As the name implies, *thermal overload relays* rely on the rising temperatures caused by the overload current to trip the overload mechanism. Thermal overload relays can be further subdivided into two types: melting alloy and bimetallic.

The *melting-alloy* assembly of heater element (overload relay) and solder pot is shown in Fig. 1-5. Excessive overload motor current passes through the heater element, thereby melting a eutectic alloy solder pot. The ratchet wheel will then be allowed to turn in the molten pool, and a tripping action of the starter control circuit results, stopping the motor. A cooling-off period is required to allow the solder pot to "freeze" before the overload relay assembly can be reset and motor service restored.

Melting-alloy thermal units are interchangeable and of one-piece construction, which ensures a constant relationship between the heater element and solder pot and allows factory calibration, making them virtually tamperproof in the field.

NEMA Size	Load Volts	Maximum Horsepower Rating — Nonplugging and Nonjogging Duty		Maximum Horsepower Rating — Plugging and Jogging Duty		Continuous Current Rating, Amperes — 600 Volt Max.	Service-Limit Current Rating, Amperes	Tungsten and Infrared Lamp Load, Amperes — 250 Volts Max.	Resistance Heating Loads, KW other than Infrared Lamp Loads		KVA Rating for Switching Transformer Primaries — Inrush (Worst Case Peak) Not More Than 20 Times Peak of Continuous Current Rating		KVA Rating for Switching Transformer Primaries — Inrush (Worst Case Peak) Over 20 Through 40 Times Peak of Continuous Current Rating		3 Phase Rating for Switching Capacitors, KVAR
		Single Phase	Poly-Phase	Single Phase	Poly-Phase				Single Phase	Poly-Phase	Single Phase	Poly-Phase	Single Phase	Poly-Phase	
00	115	1/3				9	11	5							
	200		1½			9	11	5							
	230	1	1½			9	11	5							
	380		1½			9	11								
	460		2			9	11								
	575		2			9	11								
0	115	1		½		18	21	10			0.6		0.3		
	200		3		1½	18	21	10				1.8		0.9	
	230	2	3	1	1½	18	21	10			1.2	2.1	0.6	1.0	
	380		5		1½	18	21								
	460		5		2	18	21				2.4	4.2	1.2	2.1	
	575		5		2	18	21				3.0	5.2	1.5	2.6	
1	115	2		1		27	32	15	3	5	1.2		0.6		
	200		7½		3	27	32	15		9.1		3.6		1.8	
	230	3	7½	2	3	27	32	15	6	10	2.4	4.3	1.2	2.1	
	380		10		5	27	32			16.5					
	460		10		5	27	32		12	20	4.9	8.5	2.5	4.3	
	575		10		5	27	32		15	25	6.2	11.0	3.1	5.3	
1P	115	3		1½		36	42	24							
	230	5		3		36	42	24							
2	115	3		2		45	52	30	5	8.5	2.1		1.0		
	200		10		7½	45	52	30		15.4		6.3		3.1	
	230	7½	15	5	10	45	52	30	10	17	4.1	7.2	2.1	3.6	8
	380		25		15	45	52			28					
	460		25		15	45	52		20	34	8.3	14	4.2	7.2	16
	575		25		15	45	52		25	43	10.0	18	5.2	8.9	20
3	115	7½				90	104	60	10	17	4.1		2.0		
	200		25		15	90	104	60		31		12		6.1	
	230	15	30		20	90	104	60	20	34	8.1	14	4.1	7.0	27
	380		50		30	90	104			56					
	460		50		30	90	104		40	68	16	28	8.1	14	53
	575		50		30	90	104		50	86	20	35	10	18	67

NEMA Size	Volts	Max HP (3‑Phase)	Continuous Amps	Service‑Limit Amps
4	200	40	135	156
	230	50	135	156
	380	75	135	156
	460	100	135	156
	575	100	135	156
5	200	60	270	311
	230	75	270	311
	380	125	270	311
	460	150	270	311
	575	150	270	311
6	200	125	540	621
	230	150	540	621
	380	250	540	621
	460	300	540	621
	575	300	540	621
7	230		810	932
	460		810	932
	575		810	932
8	230		1215	1400
	460		1215	1400
	575		1215	1400

Tables and footnotes are taken from NEMA Standards.

† Ratings shown are for applications requiring repeated interruptions of stalled motor current or repeated closing of high transient currents encountered in rapid motor reversal, involving more than five openings or closings per minute and more than ten in a ten‑minute period, such as plug‑stop, plug‑reverse or jogging duty. Ratings apply to single speed and multi‑speed controllers.

* Per NEMA Standards paragraph IC 1‑21A.20, the service‑limit current represents the maximum rms current, in amperes, which the controller may be expected to carry for protracted periods in normal service. At service‑limit current ratings, temperature rises may exceed those obtained by testing the controller at its continuous current rating. The ultimate trip current of over‑current (overload) relays or other motor protective devices shall not exceed the service‑limit current ratings of the controller.

★ FLUORESCENT LAMP LOADS — 300 VOLTS AND LESS — The characteristics of fluorescent lamps are such that it is not necessary to derate Class 8502 contactors below their normal continuous current rating. Class 8903 contactors may also be used for controlling tungsten and infrared lamp loads, and with fluorescent lamp loads. Class 8903 ac lighting contactors are recommended. These contactors are specifically designed for such loads and are applied at their full rating as listed in the Class 8903 Section.

‡ Ratings apply to contactors which are employed to switch the load at the utilization voltage of the heat producing element with a duty which requires continuous operation of not more than five openings per minute. Class 8903 Types L and S lighting contactors are rated for resistance heating loads.

• When discharged, a capacitor has essentially zero impedance. For repetitive switching by contactor, sufficient impedance should be connected in series to limit inrush current to not more than 6 times the contactor rated continuous current. In many installations, the impedance of connecting conductors may be sufficient for this purpose. When switching to connect additional banks, the banks already on the line may be charged and can supply additional available short‑circuit current which should be considered when selecting impedance to limit the current.

The ratings for capacitor switching above assume the following maximum available fault currents:

NEMA Size 2‑3: 5,000 A RMS Sym.
NEMA Size 4‑5: 10,000 A RMS Sym.
NEMA Size 6‑8: 22,000 A RMS Sym.

If available fault current is greater than these values, connect sufficient impedance in series as noted in the previous paragraph.

The motor ratings in the above table are NEMA standard ratings and apply only when the code letter of the motor is the same as or occurs earlier in the alphabet than is shown in the table below.

Motors having code letters occurring later in the alphabet may require a larger controller. Consult local Square D field office.

Motor HP Rating	Maximum Allowable Motor Code Letter
1½‑2	L
3‑5	K
7½ & above	H

Figure 1‑3 Electrical ratings of ac magnetic contactors and starter. (Courtesy Square D Company.)

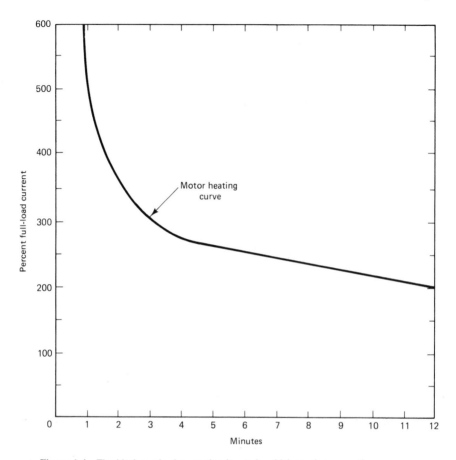

Figure 1-4 The ideal overload protection is one in which sensing properties are very similar to the heating curve of the motor.

These important features are not possible with any other type of overload relay construction. A wide selection of these interchangeable thermal units is available to give exact overload protection of any full-load current to a motor.

Bimetallic overload relays are designed specifically for two general types of application. The automatic reset feature is of decided advantage when devices are mounted in locations not easily accessible for manual operation, and these relays can easily be adjusted to trip within a range of 85 to 115% of the nominal trip rating of the heater unit. This feature is useful when the recommended heater size might result in unnecessary tripping but the next larger size would not give adequate protection. Ambient temperatures affect overload relays, operating on the principle of heat.

Ambient-compensated bimetallic overload relays were designed for one particular situation: when the motor is at a constant temperature and the controller is

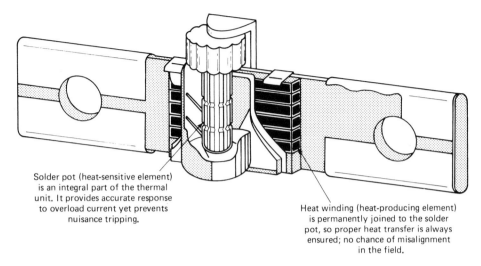

Solder pot (heat-sensitive element) is an integral part of the thermal unit. It provides accurate response to overload current yet prevents nuisance tripping.

Heat winding (heat-producing element) is permanently joined to the solder pot, so proper heat transfer is always ensured; no chance of misalignment in the field.

Figure 1-5 Melting-alloy thermal overload relay.

located separately in a varying temperature. In this case, if a standard thermal overload relay were used, it would not trip consistently at the same level of motor current if the controller temperature changed. This thermal overload relay is always affected by the surrounding temperature. To compensate for the temperature variations, the controller may see that an ambient-compensated overload relay is applied. Its trip point is not affected by temperature and it performs consistently at the same value of current.

Melting-alloy and bimetallic overload relays are designed to approximate the heat actually generated in the motor. As the motor temperature increases, so does the temperature of the thermal unit. The motor and relay heating curves (see Fig. 1-6) show this relationship. From this graph we can see that no matter how high the current is drawn, the overload relay will provide protection, yet the relay will not trip unnecessarily.

When selecting thermal overload relays, the following must be considered:

1. Motor full-load current
2. Type of motor
3. Difference in ambient temperature between motor and controller

Motors of the same horsepower rating and speed do not all have the same full-load current; the motor nameplate must always be referred to to obtain the full-load amperes for a particular motor. Do not use a published table. Thermal unit selection tables are published on the basis of continuous-duty motors with a 1.15 service factor, operating under normal conditions. The tables are shown in manufacturers' catalogs and also appear on the inside of the door or cover of the motor controller. These

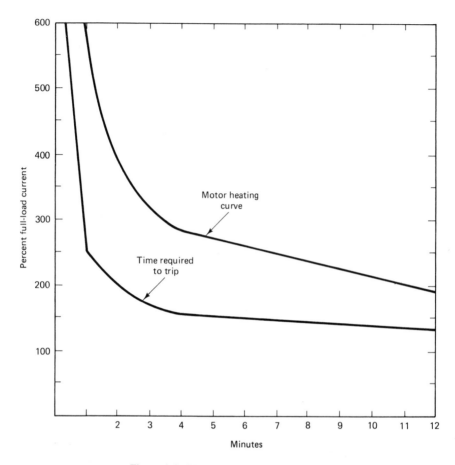

Figure 1-6   Motor and relay heating curves.

selections will protect the motor properly and allow it to develop its full horsepower rating, allowing for the service factor, if the ambient temperature is the same at the motor as at the controller. If the temperatures are not the same or if the motor service factor is less than 1.15, a special procedure is required to select the proper thermal unit.

Standard overload relay contacts are closed under normal conditions and open when the relay trips. An alarm signal is sometimes required to indicate when a motor has stopped due to an overload trip. Also, with some machines—particularly those associated with continuous processing—it may be required to signal an overload condition rather than have the motor and process stop automatically. This is done by fitting the overload relay with a set of contacts that close when the relay trips, completing the alarm circuit. These contacts are appropriately called *alarm contacts*.

    A magnetic overload relay has a movable magnetic core inside a coil which carries the motor current. The flux set up inside the coil pulls the core upward. When the core rises far enough it trips a set of contacts on the top of the relay. The movement of the core is slowed by a piston working in an oil-filled dashpot mounted below the coil. This produces an inverse-time characteristic. The effective tripping current is adjusted by moving the core on a threaded rod. The tripping time is varied by uncovering oil bypass holes in the piston. Because of the time and current adjustments, the magnetic overload relay is sometimes used to protect motors having long accelerating times or unusual duty cycles.

PROTECTIVE ENCLOSURES

The correct selection and installation of an enclosure for a particular application can contribute considerably to the length of life and trouble-free operation. To shield electrically live parts from accidental contact, some form of enclosure is always necessary. This function is usually fulfilled by a general-purpose, sheet steel cabinet. Frequently, however, dust, moisture, or explosive gases make it necessary to employ a special enclosure to protect the motor controller from corrosion or the surrounding equipment from explosion. In selecting and installing control apparatus, it is always necessary to consider carefully the conditions under which the apparatus must operate; there are many applications where a general-purpose enclosure does not afford protection.

    Underwriters' Laboratories has defined the requirements for protective enclosures according to the types of hazardous conditions, and the National Electrical Manufacturers' Association has standardized enclosures based on these requirements.

    NEMA 1—General purpose. The general-purpose enclosure is intended primarily to prevent accidental contact with the enclosed apparatus. It is suitable for general-purpose applications indoors where not exposed to unusual service conditions. A NEMA 1 enclosure serves as protection against dust and light indirect splashing but is not dust-tight.

    NEMA 3—Dusttight, raintight. This enclosure is intended to provide suitable protection against specified weather hazards. A NEMA 3 enclosure is suitable for application outdoors, such as in construction work. It is also sleet resistant.

    NEMA 3R—Rainproof, sleet resistant. This enclosure protects against interference in operation of the contained equipment due to rain, and resists damage from exposure to sleet. It is designed with conduit hubs and external mounting, as well as drainage provisions.

NEMA 4—Watertight.   A watertight enclosure is designed to meet a hose test which consists of a stream of water from a hose with a 1-in. nozzle, delivering at least 65 gallons per minute. The water is directed on the enclosure from a distance of not less than 10 ft for a period of 5 minutes. During this period, it may be directed in any one or more directions as desired. There must be no leakage of water into the enclosure under these conditions.

NEMA 4X—Watertight, corrosion resistant.   These enclosures are generally constructed similarly to NEMA 4 enclosures except they are made of a material that is highly resistant to corrosion. For this reason, they are ideal in applications such as meat-packing and chemical plants, where contaminants would ordinarily destroy a steel enclosure over a period of time.

NEMA 7—Hazardous locations, class I.   These enclosures are designed to meet the application requirements of the *National Electrical Code*® (NEC) for class 1 hazardous locations. In this type of equipment, the circuit interruption occurs in air. "Class I locations are those in which flammable gases or vapors are or may be present in the air in quantities sufficient to produce explosive or ignitible mixtures."

NEMA 9—Hazardous locations, class II.   These enclosures are designed to meet the application requirements of the NEC for class II hazardous locations. "Class ll locations are those which are hazardous because of the presence of combustible dust."

The letter or letters following the type number indicates the particular group or groups of hazardous locations (as defined in the NEC) for which the enclosure is designed. The designation is incomplete without a suffix letter or letters.

NEMA 12—Industrial use.   This type of enclosure is designed for use in those industries where it is desired to exclude such materials as dust, lint, fibers and flyings, oil seepage, or coolant seepage. There are no conduit openings or knockouts in the enclosure, and mounting is by means of flanges or mounting feet.

NEMA 13—Oiltight, dusttight.   NEMA 13 enclosures are generally made of cast iron, gasketed, or permit use in the same environments as NEMA 12 devices. The essential difference is that due to its cast housing, a conduit entry is provided as an integral part of the NEMA 13 enclosure, and mounting is by means of blind holes rather than mounting brackets.

NATIONAL ELECTRICAL CODE® REQUIREMENTS

The *National Electrical Code*® deals with the installations of equipment and is concerned primarily with safety—the prevention of injury and fire hazard to persons and property arising from the use of electricity. It is adopted on a local basis, sometimes incorporating minor changes or interpretations, as the need arises. NEC rules and provisions are enforced by governmental bodies exercising legal jurisdiction over electrical installations and used by insurance inspectors. Minimum safety standards are thus assured.

Motor control equipment is designed to meet the provisions of the NEC. Code sections applying to industrial control devices are Article 430 on motors and motor controllers and Article 500 on hazardous locations.

With minor exceptions, the NEC, together with some local codes, require a disconnect means for every motor. A combination starter consists of an across-the-line starter and a disconnect means wired together in a common enclosure. Combination starters include a blade-disconnect switch, either fusible or nonfusible, while some combination starters include a thermal–magnetic trip circuit breaker. The starter may be controlled remotely with pushbuttons, selector switches, and the like, or these devices may be installed in the cover. The single device makes a neat as well as compact electrical installation that takes little mounting space.

A combination starter provides safety for the operator, because the cover of the enclosing case is interlocked with the external operating handle of the disconnecting means. The door cannot be opened with the disconnecting means closed. With the disconnect means open, access to all parts may be had, but much less hazard is involved inasmuch as there are no readily accessible parts connected to the power line. This safety feature cannot be obtained with separately enclosed starters. In addition, the cabinet is provided with a means for padlocking the disconnect in the OFF position.

TWO-WIRE CONTROL

Figure 1-7 shows wiring diagrams for a two-wire control circuit. The control itself could be a thermostat, float switch, limit switch, or other maintained contact device to the magnetic starter. When the contacts of the control device close, they complete the coil circuit of the starter, causing it to pick up and connect the motor to the lines. When the control device contacts open, the starter is deenergized, stopping the motor.

Two-wire control provides low-voltage release but not low-voltage protection. When wired as illustrated, the starter will function automatically in response to the direction of the control device, without the attention of an operator. In this type of connection, a holding circuit interlock is not necessary.

THREE-WIRE CONTROL

A three-wire control circuit uses momentary contact, start–stop buttons, and a holding circuit interlock wired in parallel with the start button to maintain the circuit. Pressing the normally open (NO) start button completes the circuit to the

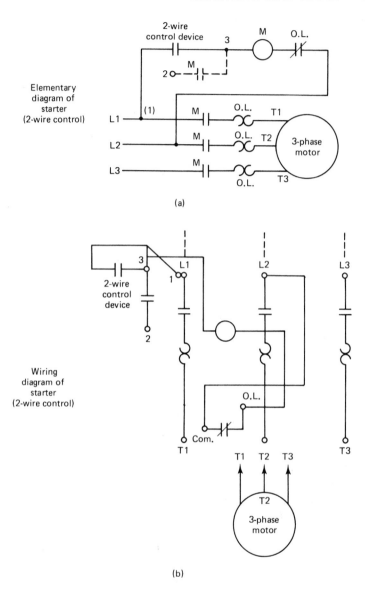

Elementary
diagram of
starter
(2-wire control)

(a)

Wiring
diagram of
starter
(2-wire control)

(b)

Figure 1-7   Elementary diagram of two-wire motor control.

coil. The power circuit contacts in lines 1, 2, and 3 close, completing the circuit to the motor, and the holding circuit contact also closes. Once the starter has picked up, the start button can be released, as the now-closed interlock contact provided an alternative current path around the reopened start contact.

Pressing the normally closed (NC) stop button will open the circuit to the coil, causing the starter to drop out. An overload condition, which caused the overload contact to open, a power failure, or a drop in voltage to less than the seal-in value would also deenergize the starter. When the starter drops out, the interlock contact reopens, and both current paths to the coil, through the start button and the interlock, are now open.

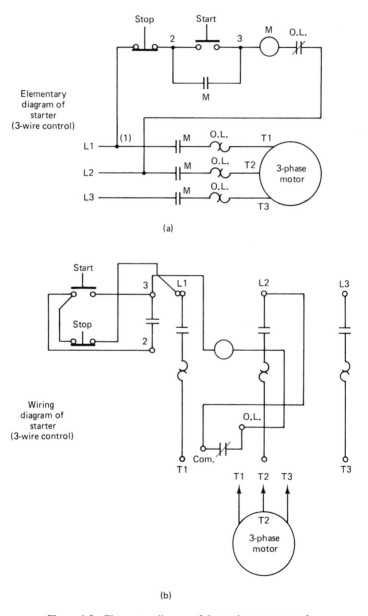

Figure 1-8 Elementary diagram of three-wire motor control.

Since three wires from the pushbutton station are connected into the starter—at points 1, 2, and 3—this wiring scheme is commonly referred to as *three-wire control* (see Fig. 1-8).

The holding circuit interlock is a normally open auxiliary contact provided on standard magnetic starters and contactors. It closes when the coil is energized to form a holding circuit for the starter after the start button has been released.

In addition to the main or power contacts, which carry the motor current, and the holding circuit interlock, a starter can be provided with externally attached auxiliary contacts, commonly called electrical interlocks. Interlocks are rated to carry only control circuit currents, not motor currents. Both NO and NC versions are available. Among a wide variety of applications, interlocks can be used to control other magnetic devices where sequence operation is desired, to electrically prevent another controller from being energized at the same time, and to make and break circuits to indicating or alarm devices such as pilot lights, bells, or other signals.

The circuit in Fig. 1-9 shows a three-pole reversing starter used to control a three-phase motor. Three-phase squirrel-cage motors can be reversed by reconnecting any two of the three-line connections to the motor. By interwiring two contactors, an electromagnetic method of making the reconnection can be obtained.

As seen in the power circuit (Fig. 1-9), the contacts (F) of the forward contactor—when closed—connect lines 1, 2, and 3 to the motor terminals T1, T2, and T3, respectively. As long as the forward contacts are closed, mechanical and electrical interlocks prevent the reverse contactor from being energized.

When the forward contactor is deenergized, the second contactor can be picked up, closing its contacts (R), which reconnect the lines to the motor. Note that by running through the reverse contacts, line 1 is connected to motor terminal T3, and line 3 is connected to motor terminal T1. The motor will now run in reverse.

Manual reversing starters (employing two manual starters) are also available. As in the magnetic version, the forward and reverse switching mechanisms are

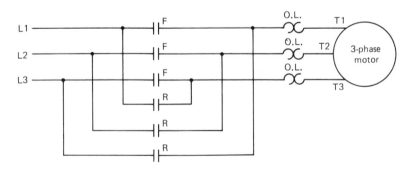

Figure 1-9 Diagram of a three-pole reversing starter used to control a three-phase motor.

mechanically interlocked, but since coils are not used in the manually operated equipment, electrical interlocks are not furnished.

CONTROL RELAYS

A relay is an electromagnetic device whose contacts are used in control circuits of magnetic starters, contactors, solenoids, timers, and other relays. They are generally used to amplify the contact capability or multiply the switching functions of a pilot device.

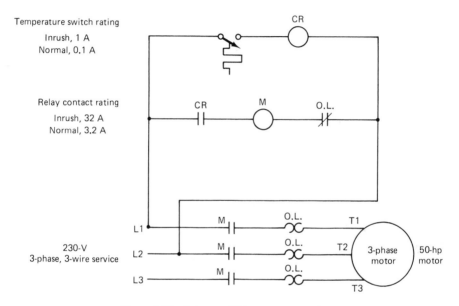

Figure 1-10 Relay amplifying contact capacity.

The wiring diagrams in Figs. 1-10 and 1-11 demonstrate how a relay amplifies contact capacity. Figure 1-10 represents current amplification. Relay and starter coil voltages are the same, but the ampere rating of the temperature switch is too low to handle the current drawn by the starter coil (M). A relay is interposed between the temperature switch and the starter coil. The current drawn by the relay coil (CR) is within the rating of the temperature switch, and relay contact (CR) has a rating adequate for the current drawn by the starter coil.

Figure 1-11 represents voltage amplification. A condition may exist in which the voltage rating of the temperature switch is too low to permit its direct use in a starter control circuit operating at a higher voltage. In this application, the coil of the interposing relay and the pilot device are wired to a low-voltage source of power

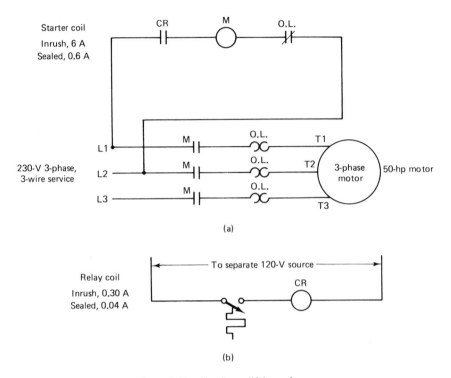

Figure 1-11   Circuit amplifying voltage.

compatible with the rating of the pilot device. The relay contact, with its higher voltage rating, is then used to control operation of the starter.

OTHER CONTROLLING EQUIPMENT

Timers and Timing Relays

A pneumatic timer or timing relay is similar to a control relay, except that certain of its contacts are designed to operate at a preset time interval after the coil is energized or deenergized. A delay on energization is also referred to as "on delay." A time delay on deenergization is also called "off delay."

A timed function is useful in applications such as the lubrication system of a large machine, in which a small oil pump must deliver lubricant to the bearings of the main motor for a set period before the main motor starts.

In pneumatic timers, the timing is accomplished by the transfer of air through a restricted orifice. The amount of restriction is controlled by an adjustable needle valve, permitting changes to be made in the timing period.

Drum Switches

A drum switch is a manually operated three-position three-pole switch which carries a horsepower rating and is used for manual reversing of single- or three-phase motors. Drum switches are available in several sizes and can be of spring-return-to-off (momentary contact) or maintained-contact type. Separate overload protection, by manual or magnetic starters, must usually be provided, as drum switches do not include this feature.

Pushbutton Stations

A control station may contain pushbuttons, selector switches, and pilot lights. Pushbuttons may be of momentary- or maintained-contact type. Selector switches are usually maintained-contact type but can be spring-return type to give momentary-contact operation.

Standard-duty stations will handle the coil currents of contactors up to size 4. Heavy-duty stations have higher contact ratings and provide greater flexibility through a wider variety of operators and interchangeability of units.

Foot Switches

A foot switch is a control device operated by a foot pedal used where the process or machine requires that the operator have both hands free. Foot switches usually have momentary contacts but are available with latches that enable them to be used as maintained contact devices.

Limit Switches

A limit switch is a control device that converts mechanical motion into an electrical control signal. Its main function is to limit movement, usually by opening a control circuit when the limit of travel is reached. Limit switches may be momentary-contact (spring-return) or maintained-contact types. Among other applications, limit switches can be used to start, stop, reverse, slow down, speed up, or recycle machine operations.

Snap Switches

Snap switches for motor control purposes are enclosed, precision switches which require low operating forces and have a high repeat accuracy. They are used as interlocks and as the switch mechanism for control devices such as precision limit switches and pressure switches. They are also available with integral operators for use as compact limit switches, door-operated interlocks, and so on. Single-pole double-throw and two-pole double-throw versions are available.

Pressure Switches

The control of pumps, air compressors, and machine tools requires control devices that respond to the pressure of a medium such as water, air, or oil. The control device that does this is a pressure switch. It has a set of contacts which are operated by the movement of a piston, bellows, or diaphragm against a set of springs. The spring pressure determines the pressures at which the switch closes and opens its contacts.

Float Switches

When a pump motor must be started and stopped according to changes in water (or other liquid) level in a tank or sump, a float switch is used. This is a control device whose contacts are controlled by movement of a rod or chain and counterweight, fitted with a float. For closed-tank applications, the movement of a float arm is transmitted through a bellows seal to the contact mechanism.

ELECTRONIC CONTROLS

In recent years, many solid-state devices and circuits have entered the motor control field—reducing previously bulky equipment to compact, efficient, and reliable electronic units. Even electric motors have been built on printed-circuit boards with all windings made from flimsy copper foil mounted on a flat card.

The Square D NORPAK^R Solid State Logic Control is one type of control system that has recently been introduced. The NOR is the basic logic element for NORPAK. Using this single element allows many functions to be obtained through the building-block approach. This results in simplicity of design and the minimum number of logic elements in a system, as well as being of great benefit when changes are required. Although NORs can be connected to form any logic function, it is sometimes more convenient to have other logic functions available. For this reason, ANDs, sealed ANDs, ORs, and memories are available. Other functions that cannot be made up from NORs alone are timers, single shots, transfer memories, counters, and shift registers. This variety of components makes this system a wise choice for any application from simple machine control to complex systems requiring counting and data manipulation. The basic design of a NORPAK system is the same whether plug-in or encapsulated components are used.

ENERGY MANAGEMENT SYSTEMS

The increasing cost of electricity and the shortage of fuel are major items of concern for management today. Utility bills have risen to levels where action must be taken to help maintain a facility's profitability.

Most commercial and industrial electric bills are made up of two charges, energy and demand. The *energy charge* is based on the quantity of energy consumed for the billing period, while in most cases, the *demand charge* is based on the peak electrical energy used during short periods of time called *demand intervals*.

One possible way to lower demand and energy costs is through the use of an *energy management system,* which is a technique of automatically controlling the demand and energy consumption of a facility to a lower and more economical level by shedding and cycling noncritical loads for brief periods. This can be accomplished by using demand controllers or, where economically feasible, by using complex computer-based systems. In some applications, the larger systems are justified, but in many cases, smaller, less expensive controllers will do the job.

With the advent of the microprocessor, the features of the demand controller together with many of the added features of computer-based systems have been combined into Square D's Class 8865 EM WATCHDOG Energy Management Systems. The controllers use a continuously integrating demand control technique, based on an electronic version of the conventional thermal legged-demand meter. Loads are shed when the predicted demand equals the programmed demand limit.

The priorities, shed and restore times, and system data can be planned out on a data-entry worksheet. A calculator-type keyboard is used to enter worksheet data into the controller. A digital readout display confirms that data have been entered.

If a programming error is made, a digital readout will display an error code. To correct an error, the clear entry key is touched and the correct data are reentered. All programming data may then be easily entered or changed.

After all data have been entered, a key switch can be locked into the RUN position, which prevents unauthorized persons from changing critical load data. While in the RUN mode, the controller's digital readout will automatically display the predicted demand, the demand limit, and the time of day. All other data settings can be displayed upon request. A battery backup is provided to retain stored information in the event of a power failure.

PROGRAMMABLE CONTROLLERS

Programmable controllers may be used in place of conventional controls; they include relays offering faster startup, decreased startup costs, quick program changes, fast troubleshooting, and up-to-date schematic diagram printouts. They also offer additional benefits, such as precision digital timing, counting, data manipulation, remote inputs and outputs, redundant control, supervisory control, process control, management report generation, machine cycle, controller self-diagnostics, dynamic graphic displays, and the like. Applications include machine tool, sequential, process, conveyor, batching, and energy management control applications.

SUMMARY

Chapter 1 has been a review of motor controllers and controlling techniques. The seasoned professional will find this information quite understandable; however, the student or budding professional may find many of the terms and explanations a bit hazy or may even be completely at a loss. The glossary in Appendix A will be of some help, but the student may need additional information before the techniques described in this chapter are completely understood. The next few chapters give the student the required basic knowledge to use the more advanced techniques covered in later chapters. It is suggested that once Chapters 1 through 5 have been read, Chapter 1 should be reread—and a much better understanding will be possible.

On the other hand, those who have been working extensively with motor controls may find the first few chapters rather elementary. Still, even the more advanced technicians and engineers should gain some value from the material presented. So, upon approaching the later chapters, there will be information of value to almost everyone involved in motor control design, application, and installation.

chapter two

Blueprint Reading

Those involved with motor controls in any capacity will encounter many types of drawings and diagrams. For example, the engineer or designer will need to study the working arrangement of many types of motor controls in order to design a suitable control system for any project that may arise. Furthermore, he or she must be able to make sketches so that draftsmen may complete working drawings for the workers who will install and maintain the system. Draftsmen must be able to read blueprints so that they can interpret the engineer's sketches, and workers on the job must be able to read drawings and diagrams so that the various connections are made correctly. Therefore, a brief sampling of the various types of drawings that may be encountered in the electric motor control field is in order.

PICTORIAL DRAWINGS

In this type of drawing the objects are drawn in one view only; that is, three-dimensional effects are simulated on the flat plane of drawing paper by drawing several faces of an object in a single view. This pictorial drawing is very useful in describing objects and conveying information to those who are not well trained in blueprint reading or in supplementing conventional diagrams in certain special cases.

One example of a pictorial drawing would be an exploded view of a motor starter to show the physical relationship of each part to the others so that the starter could be disassembled and reassembled during maintenance (see Fig. 2-1).

Three types of pictorial drawings are in common use in the electrical/electronic industry:

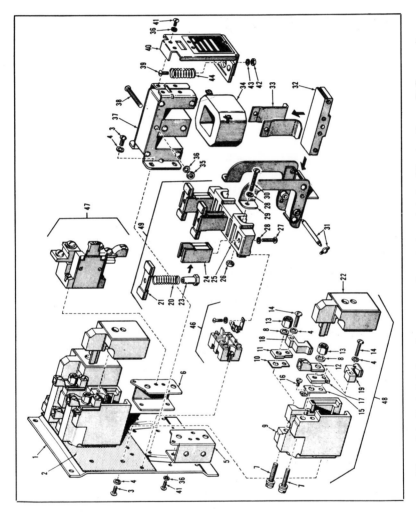

Figure 2-1 Exploded view of an ac magnetic contactor and starter. (Courtesy Square D Company.)

1. Isometric drawing
2. Oblique drawing
3. Perspective drawing

All of these drawings are relatively difficult to prepare, and they are normally used only by manufacturers of motor control components to display their products in catalogs, brochures, and similar publications. However, they are gradually being replaced by photographs where possible.

By definition, an *isometric drawing* is a view projected onto a vertical plane in which all edges are foreshortened equally. Figure 2-2 shows an isometric drawing of a cube. In this view, the edges are 120° apart and are called the *isometric axes;* the three surfaces shown are called the *isometric planes.* The lines parallel to the isometric axes are called the *isometric lines.*

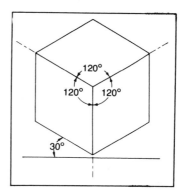

Figure 2-2   Isometric drawing of a cube.

Isometric drawings are usually preferred over oblique and perspective drawings for use in engineering departments to show certain details on installation drawings, because it is possible to draw isometric lines to scale with a 30–60° triangle.

The *oblique drawing* is similar to the isometric drawing in that one face of the object is drawn in its true shape and the other visible faces are shown by parallel lines drawn at the same angle (usually 45–30°) with the horizontal. However, unlike an isometric drawing, the lines drawn at a 30° angle are shortened to preserve the appearance of the object and are therefore not drawn to scale. Figure 2-3 shows an oblique drawing of a cube.

The two methods of pictorial drawing described so far produce only approximate representations of objects as they appear to the eye, as each type produces some degree of distortion of any object so drawn. However, because of certain advantages, these two types are the ones most often found in engineering drawings.

Sometimes—as for a certain catalog illustration or a more detailed instruction manual—it is desired to draw an exact pictorial representation of an object as it

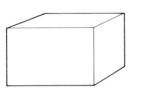

Figure 2-3 Oblique drawing of a cube.

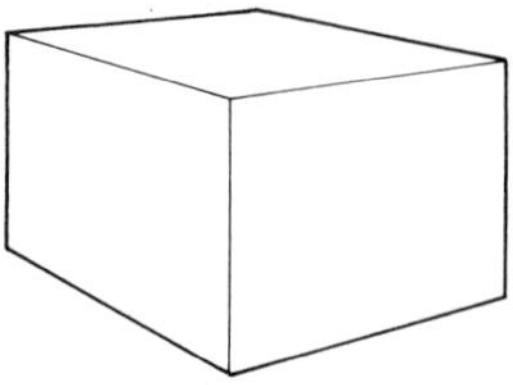

Figure 2-4 Perspective drawing of a cube.

actually appears to the eye. A drawing of this type is called a *perspective drawing;*
one such drawing—again of the cube—appears in Fig. 2-4.

ORTHOGRAPHIC-PROJECTION DRAWINGS

An orthographic-projection drawing represents the physical arrangement and views
of specific objects. These drawings give all plan views, elevation views, dimensions, and other details necessary to construct the project or object. For example,
Fig. 2-5 suggests the form of a block, but it does not show the actual shape of the
surfaces, nor does it show the dimensions of the object so that it may be
constructed.

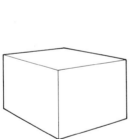

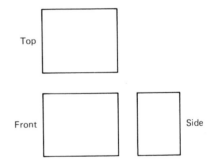

Figure 2-5 Perspective drawing suggesting the form of a
block.

Figure 2-6 Orthographic projection of the
block shown in Fig. 2-5.

An orthographic projection of the block in Fig. 2-5 is shown in Fig. 2-6. One
of the drawings in this figure shows the block as though the observer were looking
straight at the front; one, as though the observer were looking straight at the left
side; one, as though the observer were looking straight at the right side; and one, as
though the observer were looking at the rear of the block. The remaining view is as
if the observer were looking straight down on the block. These views, when com-

bined with dimensions, will allow the object to be constructed properly from materials such as metal, wood, plastic, or whatever the specifications call for.

ELECTRICAL DIAGRAMS

Electrical diagrams show, in diagrammatic form, electrical components and their related connections. Such drawings are seldom drawn to scale and show only the electrical association of the different components. In diagram drawings, symbols are used extensively to represent various pieces of electrical equipment or components, and lines are used to connect these symbols—indicating the size, type, number of wires, and the like.

In general, the types of diagrams that will be encountered by those working with motor controls will include flow diagrams (Fig. 2-7), single-line block diagrams (Fig. 2-8), and schematic wiring diagrams (Fig. 2-9).

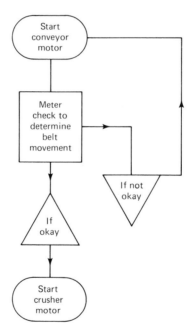

Figure 2-7   Flow diagram.

ELECTRICAL/ELECTRONIC GRAPHIC SYMBOLS

The purpose of a "working" drawing—as applied to the electrical motor control industry—is to show how a certain object, piece of equipment, or system is to be constructed, installed, modified, or repaired. An electronic testing instrument, for

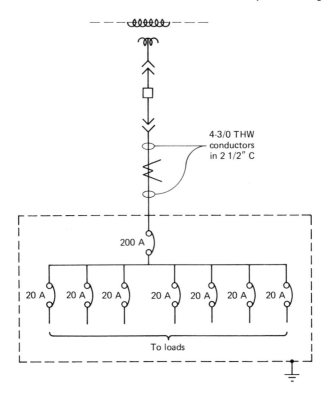

Figure 2-8   Single-line block diagram.

example, usually has drawings and specifications showing the mechanical arrangement of the chassis and housing, and a schematic diagram showing the various
components, the power supply, and the connections between each. An electrical
drawing of a system used in a building to indicate the routing of the control circuits
usually shows the floor plans of each level, the routing of the conduit or conductors,

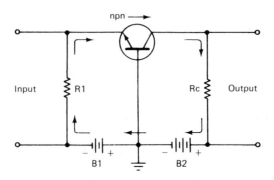

Figure 2-9   Schematic wiring diagram.

the number and sizes of wires or cables, motor control equipment, feeders, and other information for the proper installation of the system.

In preparing drawings for the electrical industry, symbols are used to simplify the work of those preparing the drawing. To illustrate this fact, look at the pictorial drawing of a motor starter in Fig. 2-10. Although this drawing clearly indicates the type of control, method of connections, and the like, such drawings would take hours for a draftsman to complete, costing more than could possibly be allotted for

Figure 2-10 Pictorial drawing of a motor starter.

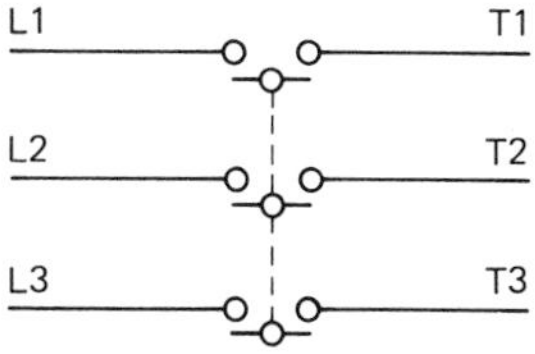

Figure 2-11 Wiring diagram of the motor starter shown in Fig. 2-10.

the conventional working drawing used in electrical construction. However, by using a drawing such as the one in Fig. 2-11, which utilizes symbols to indicate the various components, the drafting time can be cut back to minutes, and to the experienced worker, both drawings relay the exact same information.

Most engineers, designers, and draftsmen use symbols adopted by the United States of America Standards Institute (USASI) for use on electrical and electronic drawings. However, many designers and draftsmen frequently modify these symbols to suit the particular requirements of the type of work they normally encounter. For this reason, most working drawings will have a symbol list or legend placed on the drawing to describe exactly what each symbol means—eliminating practically all doubt as to what is exactly required. A typical symbol list or legend appears in Fig. 2-12.

It is evident from the list in Fig. 2-12 that many symbols have the same basic form, but their meanings differ slightly because of the addition of a line, mark, or abbreviation. Therefore, a good procedure to follow in learning the different electrical symbols is first to learn the basic form and then apply the variations of that form to obtain the different meanings.

Note also that some of the symbols in Fig. 2-12 are abbreviations, such as XFER for transfer and WT for watertight. Others are simplified pictographs, such as ☐ for externally operated disconnect switch, or NF for a nonfusible safety switch, using both pictographs and abbreviations.

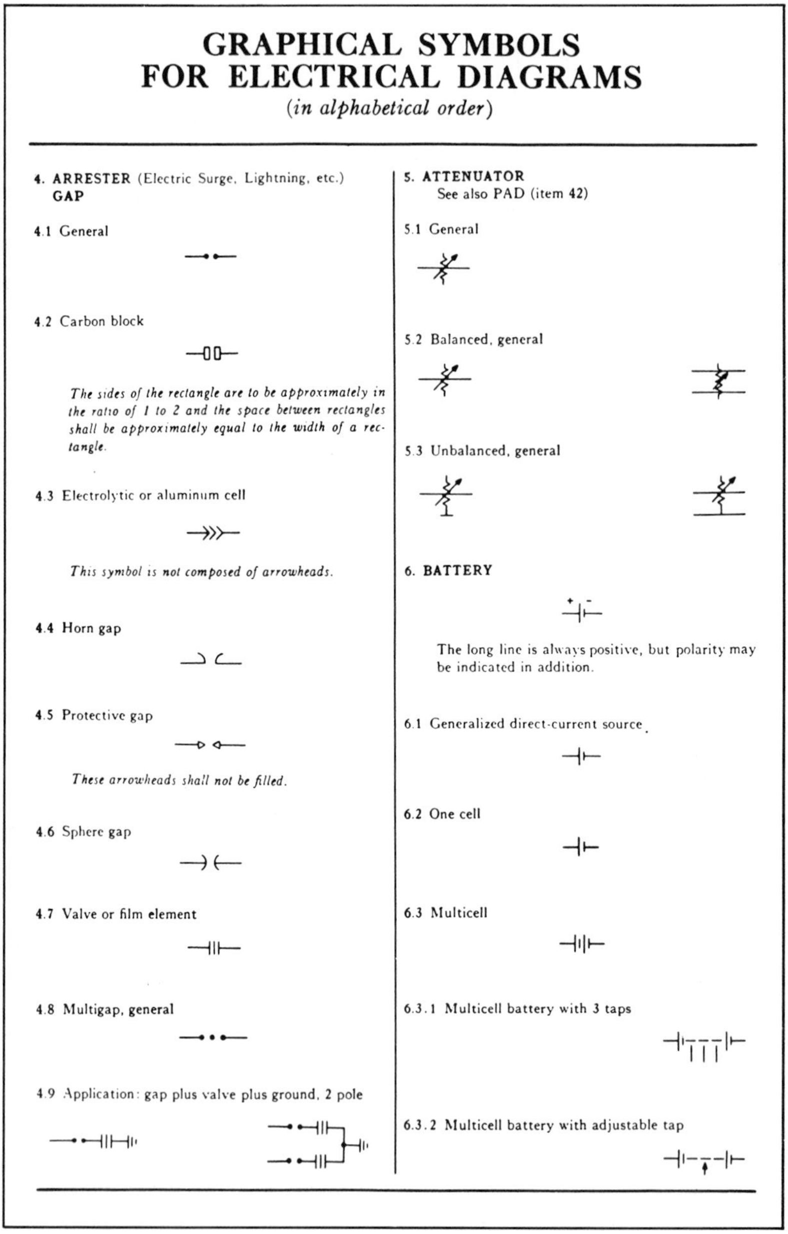

GRAPHICAL SYMBOLS
FOR ELECTRICAL DIAGRAMS
(in alphabetical order)

4. ARRESTER (Electric Surge, Lightning, etc.)
GAP

4.1 General

4.2 Carbon block

The sides of the rectangle are to be approximately in the ratio of 1 to 2 and the space between rectangles shall be approximately equal to the width of a rectangle.

4.3 Electrolytic or aluminum cell

This symbol is not composed of arrowheads.

4.4 Horn gap

4.5 Protective gap

These arrowheads shall not be filled.

4.6 Sphere gap

4.7 Valve or film element

4.8 Multigap, general

4.9 Application: gap plus valve plus ground, 2 pole

5. ATTENUATOR
See also PAD (item 42)

5.1 General

5.2 Balanced, general

5.3 Unbalanced, general

6. BATTERY

The long line is always positive, but polarity may be indicated in addition.

6.1 Generalized direct-current source

6.2 One cell

6.3 Multicell

6.3.1 Multicell battery with 3 taps

6.3.2 Multicell battery with adjustable tap

Figure 2-12 List of symbols used on electrical/electronic drawings. (Courtesy Westinghouse.)

GRAPHICAL SYMBOLS
FOR ELECTRICAL DIAGRAMS
(in alphabetical order)

7. BREAKER, CIRCUIT

If it is desired to show the condition causing the breaker to trip, the relay-protective-function symbols in item 48.8 may be used alongside the breaker symbol.

7.1 General

Use appropriate number of single-line diagram symbols

7.2 Air or, if distinction is needed, for alternating-current circuit breaker rated at 1,500 volts or less and for direct-current circuit breaker.

Use appropriate number of single-line diagram symbols

7.3 Circuit breaker, other than covered by item 7.2. The symbol in the "complete" column is for a 3-pole breaker.

On a power diagram, the symbol may be used without other identification. On a composite drawing where confusion with the general symbol (item 25) may result, add the identifying letters CB inside or adjacent to the square.

7.3.1 On a connection or wiring diagram, a 3-pole single-throw circuit breaker (with terminals shown) may be drawn as shown below.

7.4 Applications

7.4.1 3-pole circuit breaker with thermal overload device in all 3 poles.

7.4.2 3-pole circuit breaker with magnetic overload device in all 3 poles.

7.4.3 3-pole circuit breaker, drawout type

The part between the arrowheads is the movable portion.

8. CAPACITOR

See also TERMINATION (item 59.4).

8.1 General

If it is necessary to identify the capacitor electrodes, the curved element shall represent the outside electrode in fixed paper-dielectric and ceramic-dielectric capacitors, the negative electrode in electrolytic capacitors, the moving element in adjustable and variable capacitors, and the low-potential element in feed-through capacitors.

8.1.1 Application: shielded capacitor (If distinction is needed)

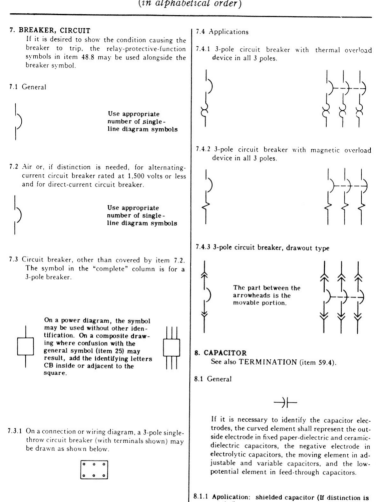

Figure 2-12 (continued)

GRAPHICAL SYMBOLS
FOR ELECTRICAL DIAGRAMS
(in alphabetical order)

8.1.2 Application: adjustable or variable capacitor

If it is necessary to identify trimmer capacitors, the letter T should appear adjacent to the symbol.

8.1.3 Application: adjustable or variable capacitors with mechanical linkage of units

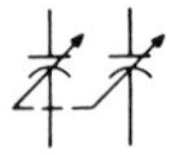

8.2 Continuously adjustable or variable differential capacitor

The capacitance of one part increases as the capacitance of the other part decreases.

8.2.1 Phase-shifter capacitor

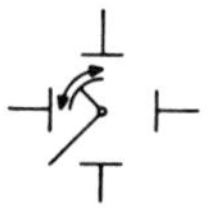

8.3 Split-stator capacitor

The capacitances of both parts increase simultaneously.

8.4 Shunt capacitor

8.5 Feed-through capacitor (with terminals shown on feed-through element)

Commonly used for bypassing high-frequency currents to chassis.

8.5.1 Application: feed-through capacitor between 2 inductors with third lead connected to chassis

8.6 Capacitance bushing for circuit breaker or transformer

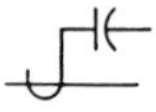

8.6.1 Application: capacitance-bushing potential device

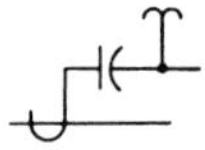

8.7 Application: coupling-capacitor potential device

9. **CELL, PHOTOSENSITIVE** (Semiconductor)

See also PHOTOTUBE (item 64.11.6).

λ indicates that the primary characteristic of the element within the circle is designed to vary under the influence of light.

9.1 Asymmetrical photoconductive transducer (resistive)

The arrowhead shall be solid

Figure 2-12 (continued)

GRAPHICAL SYMBOLS
FOR ELECTRICAL DIAGRAMS
(in alphabetical order)

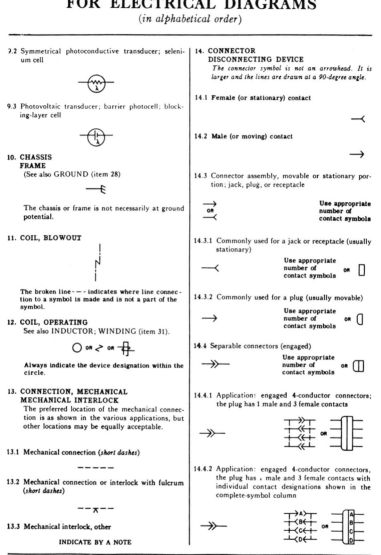

9.2 Symmetrical photoconductive transducer; selenium cell

9.3 Photovoltaic transducer; barrier photocell; blocking-layer cell

10. **CHASSIS**
    FRAME
    (See also GROUND (item 28))

    The chassis or frame is not necessarily at ground potential.

11. **COIL, BLOWOUT**

    The broken line - — - indicates where line connection to a symbol is made and is not a part of the symbol.

12. **COIL, OPERATING**
    See also INDUCTOR; WINDING (item 31).

    Always indicate the device designation within the circle.

13. **CONNECTION, MECHANICAL**
    MECHANICAL INTERLOCK
    The preferred location of the mechanical connection is as shown in the various applications, but other locations may be equally acceptable.

13.1 Mechanical connection *(short dashes)*

13.2 Mechanical connection or interlock with fulcrum *(short dashes)*

13.3 Mechanical interlock, other

    INDICATE BY A NOTE

14. **CONNECTOR**
    DISCONNECTING DEVICE
    The connector symbol is not an arrowhead. It is larger and the lines are drawn at a 90-degree angle.

14.1 **Female (or stationary) contact**

14.2 **Male (or moving) contact**

14.3 Connector assembly, movable or stationary portion; jack, plug, or receptacle

    OR   Use appropriate
                                              number of
                                              contact symbols

14.3.1 Commonly used for a jack or receptacle (usually stationary)

    Use appropriate
    number of   **OR**
    contact symbols

14.3.2 Commonly used for a plug (usually movable)

    Use appropriate
    number of   **OR**
    contact symbols

14.4 Separable connectors (engaged)

    Use appropriate
    number of   **OR**
    contact symbols

14.4.1 Application: engaged 4-conductor connectors; the plug has 1 male and 3 female contacts

14.4.2 Application: engaged 4-conductor connectors, the plug has 1 male and 3 female contacts with individual contact designations shown in the complete-symbol column

Figure 2-12 (continued)

GRAPHICAL SYMBOLS
FOR ELECTRICAL DIAGRAMS
(in alphabetical order)

14.5 Coaxial connectors

14.5.1 Engaged coaxial connectors

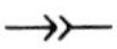

Coaxial recognition sign may be added if necessary. See PATH, TRANSMISSION (items 43.1 and 43.8.2).

14.5.1.1 If it is necessary to show that the outside conductor is carried through

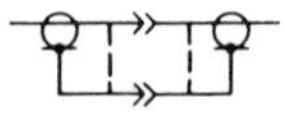

14.5.1.2 If coaxial is connected to a single conductor

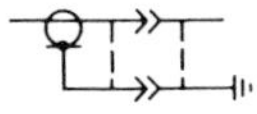

14.6 Communication switchboard-type connector

14.6.1 2-conductor (jack)

14.6.2 2-conductor (plug)

14.6.3 3-conductor (jack) with 2 break contacts (normals) and 1 auxiliary make contact

14.6.4 3-conductor (plug)

14.7 Communication switchboard-type connector with circuit normalled through

"Normalled" indicates that a through circuit may be interrupted by an inserted connector. As shown here, the inserted connector opens the through circuit and connects to the circuit towards the left.

Items 14.7.1 through 14.7.4 show 2-conductor jacks. The "normal" symbol is applicable to other types of connectors.

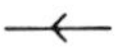

14.7.1 Jacks with circuit normalled through one way

14.7.2 Jacks with circuit normalled through both ways

14.7.3 Jacks in multiple, one set with circuit normalled through both ways

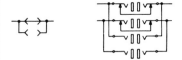

14.7.4 Jacks with auxiliary contacts, with circuit normalled through both ways

14.8 Connectors of the type commonly used for power-supply purposes (convenience outlets and mating connectors)

14.8.1 Female contact

Figure 2-12 (continued)

GRAPHICAL SYMBOLS
FOR ELECTRICAL DIAGRAMS
(in alphabetical order)

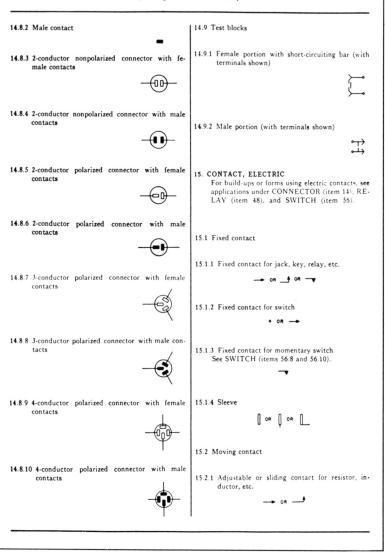

14.8.2 Male contact

14.8.3 2-conductor nonpolarized connector with female contacts

14.8.4 2-conductor nonpolarized connector with male contacts

14.8.5 2-conductor polarized connector with female contacts

14.8.6 2-conductor polarized connector with male contacts

14.8.7 3-conductor polarized connector with female contacts

14.8.8 3-conductor polarized connector with male contacts

14.8.9 4-conductor polarized connector with female contacts

14.8.10 4-conductor polarized connector with male contacts

14.9 Test blocks

14.9.1 Female portion with short-circuiting bar (with terminals shown)

14.9.2 Male portion (with terminals shown)

15. CONTACT, ELECTRIC
For build-ups or forms using electric contacts, see applications under CONNECTOR (item 14), RELAY (item 48), and SWITCH (item 56).

15.1 Fixed contact

15.1.1 Fixed contact for jack, key, relay, etc.

15.1.2 Fixed contact for switch

15.1.3 Fixed contact for momentary switch
See SWITCH (items 56.8 and 56.10).

15.1.4 Sleeve

15.2 Moving contact

15.2.1 Adjustable or sliding contact for resistor, inductor, etc.

Figure 2-12 (continued)

GRAPHICAL SYMBOLS
FOR ELECTRICAL DIAGRAMS
(in alphabetical order)

15.2.2 Locking

15.2.3 Nonlocking

15.2.4 Segment; bridging contact
   See SWITCH (items 56.12.3 and 56.12.4).

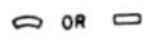

   ⌒ OR ⌐

15.2.5 Vibrator reed

15.2.6 Vibrator split reed

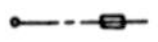

15.2.7 Rotating contact (slip ring) and brush

15.3 Basic contact assemblies

The standard method of showing a contact is by
a symbol indicating the circuit condition it produces when the actuating device is in the deenergized or nonoperated position. The actuating
device may be of a mechanical, electrical, or other
nature, and a clarifying note may be necessary with
the symbol to explain the proper point at which
the contact functions, for example, the point
where a contact closes or opens as a function of
changing pressure, level, flow, voltage, current,
etc. In cases where it is desirable to show contacts
in the energized or operated condition and where
confusion may result, a clarifying note shall be
added to the drawing.

Auxiliary switches or contacts for circuit breakers,
safety enclosed trucks, removable circuit-breaker
units, housings, enclosures, etc., may be designated as follows:
(a) Closed when device is in energized or operated
    position,
(b) Closed when device is in de-energized or non-
    operated position,
(aa) Closed when operating mechanism of main
    device is in energized or operated position,
(bb) Closed when operating mechanism of main
    device is in de-energized or nonoperated posi-
    tion.

As applied to a removable circuit-breaker unit,
(a) is an auxiliary contact that is closed when the
unit is in the connected position. As applied to a
housing or enclosure, (a) is an auxiliary contact
that is closed when the removable circuit-breaker
unit is in the connected position. See latest issue of
American Standard C37.2 for further details.

*In the parallel-line contact symbols shown below, the
length of the parallel lines shall be approximately $1\frac{1}{2}$
times the width of the gap (except for item 15.6).*

15.3.1 Closed contact (break)

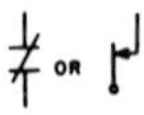

15.3.2 Open contact (make)

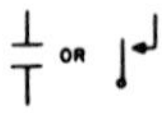

15.3.3 Transfer

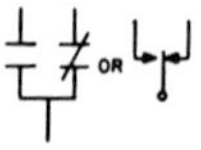

15.3.4 Make-before-break

15.4 Application: open contact with time closing (TC
   or TDC) feature

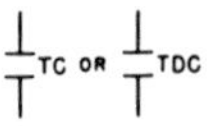

15.5 Application: closed contact with time opening
   (TO or TDO) feature

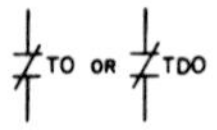

15.6 Time sequential closing

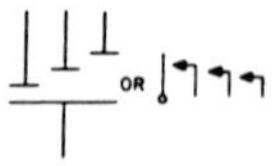

Figure 2-12 (continued)

GRAPHICAL SYMBOLS
FOR ELECTRICAL DIAGRAMS
(in alphabetical order)

16. CONTACTOR

See also RELAY (item 48)

Fundamental symbols for contacts, coils, mechanical connections, etc., are the basis of contactor symbols and should be used to represent contactors on complete diagrams. Complete diagrams of contactors consist of combinations of fundamental symbols for control coils, mechanical connections, etc., in such configurations as to represent the actual device.

Mechanical interlocking should be indicated by notes.

16.1 Manually operated 3-pole contactor

16.2 Electrically operated 1-pole contactor with series blowout coil

*See note below

16.3 Electrically operated 3-pole contactor with series blowout coils; 2 open and 1 closed auxiliary contacts (shown smaller than the main contacts)

*See note below   OR

16.4 Electrically operated 1-pole contactor with shunt blowout coil

*See note below

*Always indicate the device designation within the circle.

17. CORE

17.1 General or air core

NO SYMBOL

If it is necessary to identify an air core, a note should appear adjacent to the symbol of the inductor or transformer.

17.2 Magnetic core of inductor or transformer

Not to be used unless it is necessary to identify a magnetic core.

See INDUCTOR (item 31.2) and TRANSFORMER (item 63.2).

17.3 Core of magnet or relay

For use if representation of the core is necessary.

See MAGNET, PERMANENT (item 36) and RELAY (items 48.2 to 48.4 and 48.6, 48.7).

18. COUNTER, ELECTROMAGNETICALLY OPERATED
MESSAGE REGISTER

18.1 General

18.2 With a make contact

19. COUPLER, DIRECTIONAL

Commonly used in coaxial and waveguide diagrams.

The arrows indicate the direction of power flow.

Number of coupling paths, type of coupling, and transmission loss may be indicated.

19.1 General

Figure 2-12 (continued)

GRAPHICAL SYMBOLS
FOR ELECTRICAL DIAGRAMS
(in alphabetical order)

19.2 Applications

19.2.1 *E*-plane aperture coupling, 30-decibel transmission loss

19.2.2 Loop coupling, 30-decibel transmission loss

19.2.3 Probe coupling, 30-decibel transmission loss

19.2.4 Resistance coupling, 30-decibel transmission loss

20. COUPLING

Commonly used in coaxial and waveguide diagrams.

20.1 Coupling by aperture with an opening of less than full waveguide size

Always indicate the type of coupling by designation: E, H or HE within the circle.

E indicates that the physical plane of the aperture is perpendicular to the transverse component of the major *E* lines.

H indicates that the physical plane of the aperture is parallel to the transverse component of the major *E* lines.

HE indicates coupling by all other kinds of apertures.

Transmission loss may be indicated.

20.1.1 Application: *E*-plane coupling by aperture to space

20.1.2 Application: *E*-plane coupling by aperture; 2 ends of transmission path available

20.1.3 Application: *E*-plane coupling by aperture; 3 ends of transmission path available

20.1.4 Application: *E*-plane coupling by aperture; 4 ends of transmission path available

20.2 Coupling by loop to space

20.2.1 Coupling by loop to guided transmission path

20.2.2 Application: coupling by loop from coaxial to circular waveguide with direct-current grounds connected

20.3 Coupling by probe to space
See OPEN CIRCUIT (item 59.2).

20.3.1 Application: coupling by probe to a guided transmission path

20.3.2 Application: coupling by probe from coaxial to rectangular waveguide with direct-current grounds connected

Figure 2-12 (continued)

GRAPHICAL SYMBOLS
FOR ELECTRICAL DIAGRAMS
(in alphabetical order)

21. DEVICE, AUDIBLE SIGNALING

21.1 Bell, general; telephone ringer

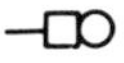

If specific identification is required, the abbreviation AC or DC may be added within the square.

21.2 Buzzer

If specific identification is required, the abbreviation AC or DC may be added within the square.

21.3 Horn; howler; loudspeaker; siren

21.3.1 General

21.3.2 If specific identification of loudspeaker parts is required, the following letter combinations may be added. The * and ‡ are not part of the symbol.

*HN   Horn
*HW   Howler
*LS   Loudspeaker
*SN   Siren
‡EM   Electromagnetic with moving coil (moving coil leads should be identified)
‡EMN   Electromagnetic with moving coil and neutralizing winding (moving coil leads should be identified)
‡MG   Magnetic armature
‡PM   Permanent magnet with moving coil

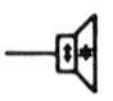

OR

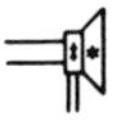

22. DEVICE, VISUAL SIGNALING

22.1 Annunciator, general

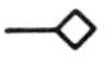

OR

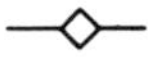

22.1.1 Annunciator drop or signal, shutter or grid type

22.1.2 Annunciator drop or signal, ball type

22.1.3 Manually restored drop

22.1.4 Electrically restored drop

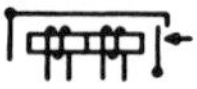

22.3 Indicating, pilot, signaling, or switchboard light

See also GLOW LAMP (item 33.3).

O   OR   ☐   OR   ⊜

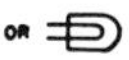

Always add the letter or letters specified below within or adjacent to the symbol. The suffix L or IL may be added to the letter or letters below to avoid confusion of the circular symbol with meter or basic relay symbols; for example, RL or RIL placed within or adjacent to the circle.

A Amber   G Green   OP Opalescent   W White
B Blue   NE Neon   P Purple   Y Yellow
C Clear   O Orange R Red

The D-shaped symbol is sometimes used to avoid confusion with other circular symbols.

Figure 2-12 (continued)

GRAPHICAL SYMBOLS
FOR ELECTRICAL DIAGRAMS
(in alphabetical order)

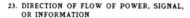

23. DIRECTION OF FLOW OF POWER, SIGNAL, OR INFORMATION

The lower symbol of each group below is used if it is necessary to conserve space. The arrowhead in the lower symbol shall be filled.

23.1 One-way

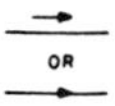

23.2 Both ways

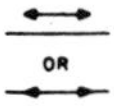

23.3 Application: one-way circuit element, general

Always indicate the type of apparatus by appropriate words or letters in the rectangle.

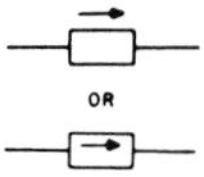

24. DISCONTINUITY

A component that exhibits throughout the frequency range of interest the properties of the type of circuit element indicated by the symbol within the triangle.

Commonly used for coaxial and waveguide transmission.

24.1 Equivalent series element, general

24.1.1 Capacitive reactance

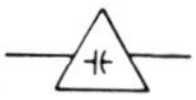

24.1.2 Inductive reactance

24.1.3 Inductance-capacitance circuit with infinite reactance at resonance

24.1.4 Inductance-capacitance circuit with zero reactance at resonance

24.1.5 Resistance

24.2 Equivalent shunt element, general

24.2.1 Capacitive susceptance

24.2.2 Conductance

24.2.3 Inductive susceptance

24.2.4 Inductance-capacitance circuit with infinite susceptance at resonance

24.2.5 Inductance-capacitance circuit with zero susceptance at resonance

Figure 2-12 (continued)

GRAPHICAL SYMBOLS
FOR ELECTRICAL DIAGRAMS
(in alphabetical order)

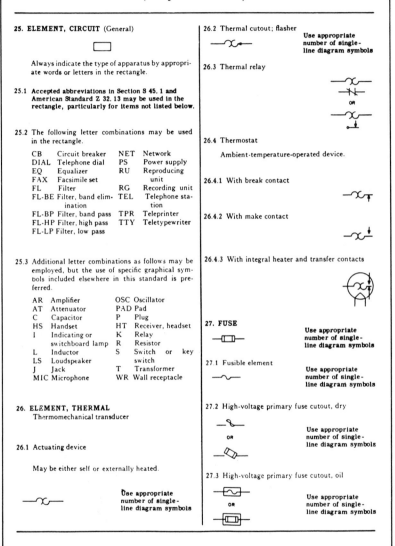

25. ELEMENT, CIRCUIT (General)

Always indicate the type of apparatus by appropriate words or letters in the rectangle.

25.1 **Accepted abbreviations in Section S 45.1 and American Standard Z 32.13 may be used in the rectangle, particularly for items not listed below.**

25.2 The following letter combinations may be used in the rectangle.

CB	Circuit breaker	NET	Network
DIAL	Telephone dial	PS	Power supply
EQ	Equalizer	RU	Reproducing
FAX	Facsimile set		unit
FL	Filter	RG	Recording unit
FL-BE	Filter, band elimination	TEL	Telephone station
FL-BP	Filter, band pass	TPR	Teleprinter
FL-HP	Filter, high pass	TTY	Teletypewriter
FL-LP	Filter, low pass		

25.3 Additional letter combinations as follows may be employed, but the use of specific graphical symbols included elsewhere in this standard is preferred.

AR	Amplifier	OSC	Oscillator
AT	Attenuator	PAD	Pad
C	Capacitor	P	Plug
HS	Handset	HT	Receiver, headset
I	Indicating or switchboard lamp	K	Relay
		R	Resistor
L	Inductor	S	Switch or key switch
LS	Loudspeaker		
J	Jack	T	Transformer
MIC	Microphone	WR	Wall receptacle

26. ELEMENT, THERMAL
Thermomechanical transducer

26.1 Actuating device

May be either self or externally heated.

Use appropriate number of single-line diagram symbols

26.2 Thermal cutout; flasher

Use appropriate number of single-line diagram symbols

26.3 Thermal relay

OR

26.4 Thermostat

Ambient-temperature-operated device.

26.4.1 With break contact

26.4.2 With make contact

26.4.3 With integral heater and transfer contacts

27. FUSE

Use appropriate number of single-line diagram symbols

27.1 Fusible element

Use appropriate number of single-line diagram symbols

27.2 High-voltage primary fuse cutout, dry

OR

Use appropriate number of single-line diagram symbols

27.3 High-voltage primary fuse cutout, oil

OR

Use appropriate number of single-line diagram symbols

Figure 2-12 (continued)

GRAPHICAL SYMBOLS
FOR ELECTRICAL DIAGRAMS
(in alphabetical order)

27.4 With alarm contact

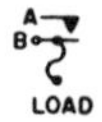

When fuse blows, alarm bus A is connected to power bus B. Letters are for explanation and are not part of the symbol.

28. GROUND
See also CHASSIS; FRAME (item 10).

29. HANDSET
OPERATOR'S SET

29.1 General

29.2 With push-to-talk switch

29.3 3-conductor handset

29.4 4-conductor handset

29.5 4-conductor handset with push-to-talk switch

29.6 Operator's set

30. HYBRID

30.1 Hybrid, general

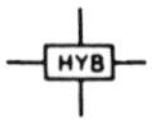

30.2 Hybrid, junction
Commonly used in coaxial and waveguide transmission.

30.3 Application: rectangular waveguide and coaxial coupling

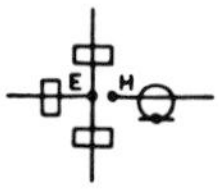

30.4 Hybrid, circular (basic)

Always place E, H, or HE within the circle. E indicates that there is a principal E transverse field in the plane of the ring. H indicates that there is a principal H transverse field in the plane of the ring. HE shall be used for all other cases.

An arm that has coupling of a different type from that designated above shall be marked according to COUPLING (item 20. 1).

Critical distances should be labeled in terms of guide wavelengths.

30.4.1 Application: 5-arm circular hybrid with principal coupling in the *E* plane and with 1-arm *H* coupling using rectangular waveguide

---

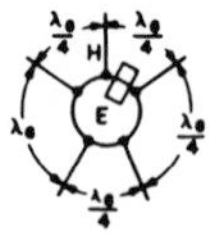

Figure 2-12 (continued)

GRAPHICAL SYMBOLS
FOR ELECTRICAL DIAGRAMS
(in alphabetical order)

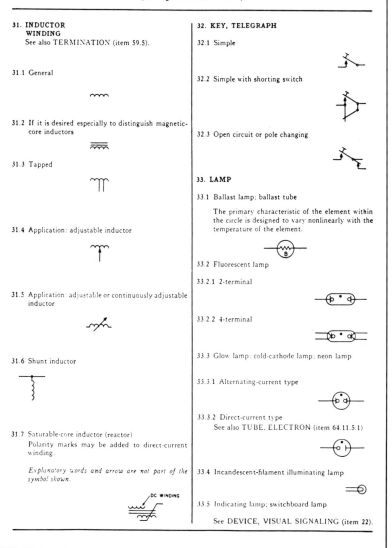

31. INDUCTOR
WINDING
See also TERMINATION (item 59.5).

31.1 General

31.2 If it is desired especially to distinguish magnetic-core inductors

31.3 Tapped

31.4 Application: adjustable inductor

31.5 Application: adjustable or continuously adjustable inductor

31.6 Shunt inductor

31.7 Saturable-core inductor (reactor)
Polarity marks may be added to direct-current winding.

Explanatory words and arrow are not part of the symbol shown.

DC WINDING

32. KEY, TELEGRAPH

32.1 Simple

32.2 Simple with shorting switch

32.3 Open circuit or pole changing

33. LAMP

33.1 Ballast lamp; ballast tube

The primary characteristic of the element within the circle is designed to vary nonlinearly with the temperature of the element.

33.2 Fluorescent lamp

33.2.1 2-terminal

33.2.2 4-terminal

33.3 Glow lamp; cold-cathode lamp; neon lamp

33.3.1 Alternating-current type

33.3.2 Direct-current type
See also TUBE, ELECTRON (item 64.11.5.1)

33.4 Incandescent-filament illuminating lamp

33.5 Indicating lamp; switchboard lamp

See DEVICE, VISUAL SIGNALING (item 22).

Figure 2-12 (continued)

GRAPHICAL SYMBOLS
FOR ELECTRICAL DIAGRAMS
(in alphabetical order)

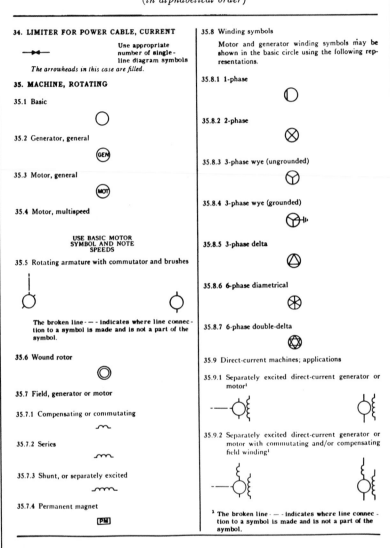

34. LIMITER FOR POWER CABLE, CURRENT

Use appropriate number of single-line diagram symbols

The arrowheads in this case are filled.

35. MACHINE, ROTATING

35.1 Basic

35.2 Generator, general

35.3 Motor, general

35.4 Motor, multispeed

USE BASIC MOTOR SYMBOL AND NOTE SPEEDS

35.5 Rotating armature with commutator and brushes

The broken line - — - indicates where line connection to a symbol is made and is not a part of the symbol.

35.6 Wound rotor

35.7 Field, generator or motor

35.7.1 Compensating or commutating

35.7.2 Series

35.7.3 Shunt, or separately excited

35.7.4 Permanent magnet

35.8 Winding symbols

Motor and generator winding symbols may be shown in the basic circle using the following representations.

35.8.1 1-phase

35.8.2 2-phase

35.8.3 3-phase wye (ungrounded)

35.8.4 3-phase wye (grounded)

35.8.5 3-phase delta

35.8.6 6-phase diametrical

35.8.7 6-phase double-delta

35.9 Direct-current machines; applications

35.9.1 Separately excited direct-current generator or motor[1]

35.9.2 Separately excited direct-current generator or motor with commutating and/or compensating field winding[1]

[1] The broken line - — - indicates where line connection to a symbol is made and is not a part of the symbol.

Figure 2-12 (continued)

GRAPHICAL SYMBOLS
FOR ELECTRICAL DIAGRAMS
(in alphabetical order)

35.9.3 Compositely excited direct-current generator or motor with commutating and/or compensating field winding[1]

35.9.4 Direct-current series motor or 2-wire generator[1]

35.9.5 Direct-current series motor or 2-wire generator with commutating and/or compensating field winding[1]

35.9.6 Direct-current shunt motor or 2-wire generator[1]

35.9.7 Direct-current shunt motor or 2-wire generator with commutating and/or compensating field winding[1]

35.9.8 Direct-current permanent-magnet-field generator or motor[1]

35.9.9 Direct-current compound motor or 2-wire generator or stabilized shunt motor[1]

35.9.10 Direct-current compound motor or 2-wire generator or stabilized shunt motor with commutating and/or compensating field winding[1]

35.9.11 Direct-current 3-wire shunt generator[1]

35.9.12 Direct-current 3-wire shunt generator with commutating and/or compensating field winding[1]

35.9.13 Direct-current 3-wire compound generator[1]

[1] The broken line - — - indicates where line connection to a symbol is made and is not a part of the symbol.

Figure 2-12 (continued)

GRAPHICAL SYMBOLS
FOR ELECTRICAL DIAGRAMS
(in alphabetical order)

35.9.14 Direct-current 3-wire compound generator with commutating and/or compensating field winding[1]

35.9.15 Direct-current balancer, shunt wound[1]

35.9.16 Direct-current balancer, compound wound[1]

35.9.17 Dynamotor[1]

35.9.18 Double-current generator[1]

35.9.19 Acyclic generator (separately excited)[1]

35.9.20 Regulating generator (rotary amplifier) shunt wound with short-circuited brushes[1]

35.9.21 Regulating generator (rotary amplifier) shunt wound without short-circuited brushes[1]

35.9.22 Regulating generator (rotary amplifier) shunt wound with compensating field winding and short-circuited brushes[1]

35.9.23 Regulating generator (rotary amplifier) shunt wound with compensating field winding but without short-circuited brushes[1]

35.10 Alternating-current machines; applications

35.10.1 Squirrel-cage induction motor or generator, split-phase induction motor or generator, rotary phase converter, or repulsion motor[1]

[1] The broken line - — - indicates where line connection to a symbol is made and is not a part of the symbol.

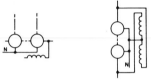

Figure 2-12 (continued)

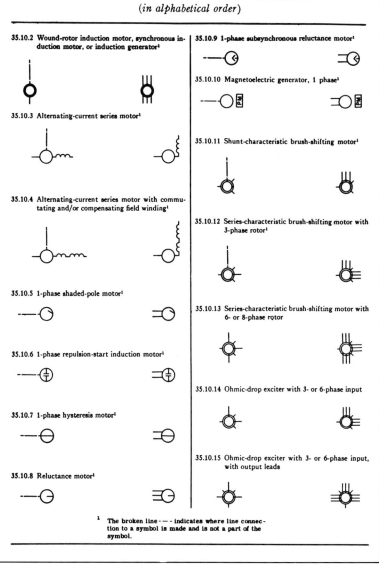

GRAPHICAL SYMBOLS
FOR ELECTRICAL DIAGRAMS
(in alphabetical order)

35.10.2 Wound-rotor induction motor, synchronous induction motor, or induction generator[1]

35.10.3 Alternating-current series motor[1]

35.10.4 Alternating-current series motor with commutating and/or compensating field winding[1]

35.10.5 1-phase shaded-pole motor[1]

35.10.6 1-phase repulsion-start induction motor[1]

35.10.7 1-phase hysteresis motor[1]

35.10.8 Reluctance motor[1]

35.10.9 1-phase subsynchronous reluctance motor[1]

35.10.10 Magnetoelectric generator, 1 phase[1]

35.10.11 Shunt-characteristic brush-shifting motor[1]

35.10.12 Series-characteristic brush-shifting motor with 3-phase rotor[1]

35.10.13 Series-characteristic brush-shifting motor with 6- or 8-phase rotor

35.10.14 Ohmic-drop exciter with 3- or 6-phase input

35.10.15 Ohmic-drop exciter with 3- or 6-phase input, with output leads

[1] The broken line - — - indicates where line connection to a symbol is made and is not a part of the symbol.

Figure 2-12 (continued)

GRAPHICAL SYMBOLS
FOR ELECTRICAL DIAGRAMS
(in alphabetical order)

35.10.16 3-phase regulating machine

35.10.17 Phase shifter with 1-phase output
See SHIFTER, PHASE (item 53).
See TRANSFORMER (item 63).

35.10.18 Phase shifter with 3-phase output
See SHIFTER, PHASE (item 53).
See TRANSFORMER (item 63).

35.11 Alternating-current machines with direct-current
field excitation; applications

35.11.1 Synchronous motor, generator, or condenser[1]

35.11.2 Synchronous motor, generator, or condenser
with neutral brought out[1]

35.11.3 Synchronous motor, generator, or condenser
with both ends of each phase brought out[1]

35.11.4 Double-winding synchronous generator, motor,
or condenser[1]

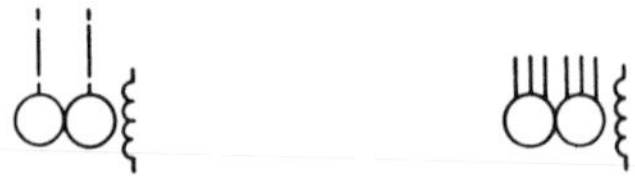

35.11.5 Synchronous-synchronous frequency changer[1]

35.11.6 Synchronous induction frequency changer[1]

35.12 Alternating- and direct-current composite machines; applications

35.12.1 Synchronous or regulating-pole converter[1]

[1] The broken line - — - indicates where line connection to a symbol is made and is not a part of the symbol.

Figure 2-12 (continued)

GRAPHICAL SYMBOLS
FOR ELECTRICAL DIAGRAMS
(in alphabetical order)

35.12.2 Synchronous booster or regulating-pole converter with commutating and/or compensating field windings[1]

35.12.3 Synchronous shunt-wound converter with commutating and/or compensating windings[1]

35.12.4 Synchronous converter compound wound with commutating and/or compensating field windings[1]

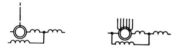

35.12.5 Motor converter[1]

36. MAGNET, PERMANENT

[1] The broken line - — - indicates where line connection to a symbol is made and is not a part of the symbol.

37. METER
INSTRUMENT

Note 17—The asterisk is not a part of the symbol. Always replace the asterisk by one of the following letter combinations, depending on the function of the meter or instrument, unless some other identification is provided in the circle and explained on the diagram.

A-	Ammeter
AH	Ampere-hour meter
CMA	Contact-making (or breaking) ammeter
CMC	Contact-making (or breaking) clock
CMV	Contact-making (or breaking) voltmeter
CRO	Oscilloscope or cathode-ray oscillograph
D	Demand meter
DB	DB (decibel) meter
DBM	DBM (decibels referred to 1 milliwatt) meter
DTR	Demand-totalizing relay
F	Frequency meter
G	Galvanometer
GD	Ground detector
I	Indicating
M	Integrating
μA or UA	Microammeter
MA	Milliammeter
N	Noise meter
OHM	Ohmmeter
OP	Oil pressure
OSCG	Oscillograph, string
PH	Phase meter
PI	Position indicator
PF	Power-factor meter
RD	Recording demand meter
REC	Recording
RF	Reactive-factor meter
S	Synchroscope
TLM	Telemeter
T	Temperature meter
TT	Total time
VH	Varhour meter
V	Voltmeter
VA	Volt-ammeter
VAR	Varmeter
VI	Volume indicator
VU	Standard volume indicator
W	Wattmeter
WH	Watthour meter

Figure 2-12 (continued)

GRAPHICAL SYMBOLS
FOR ELECTRICAL DIAGRAMS
(in alphabetical order)

38. MICROPHONE

39. MOTION, MECHANICAL

39.1 Translation, one direction

39.2 Translation, both directions

39.3 Rotation, one direction

39.4 Rotation, both directions

39.5 Rotation designation (applied to a resistor)

CW indicates position of adjustable contact at the limit of clockwise travel viewed from knob or actuator end unless otherwise indicated.

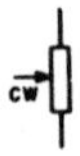

Always add identification within or adjacent to the rectangle.

For Electronics Application

40. NETWORK

40.1 General

NET

40.2 Network, low-voltage power

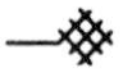

41. OSCILLATOR
GENERALIZED   ALTERNATING-CURRENT
SOURCE

42. PAD

See also ATTENUATOR (item 5)

42.1 General

42.2 Balanced, general

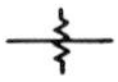

42.3 Unbalanced, general

43. PATH, TRANSMISSION
CONDUCTOR
CABLE
WIRING

43.1 Guided path, general

A single line represents the entire group of conductors or the transmission path needed to guide the power or the signal. For coaxial and waveguide work, the recognition symbol is used at the beginning and end of each kind of transmission path and at intermediate points as needed for clarity. In waveguide work, mode may be indicated.

43.2 Conductive path or conductor; wire

43.3 Air or space path

43.4 Dielectric path other than air

Commonly used for coaxial and waveguide transmission.

DIEL

Figure 2-12 (continued)

GRAPHICAL SYMBOLS
FOR ELECTRICAL DIAGRAMS
(in alphabetical order)

43.5 Crossing of paths or conductors not connected
The crossing is not necessarily at a 90-degree angle.

43.6 Junction of paths or conductors

43.6.1 Junction (if desired)

43.6.1.1 Application: junction of different-size cables

43.6.2 Junction of connected paths, conductors, or wires

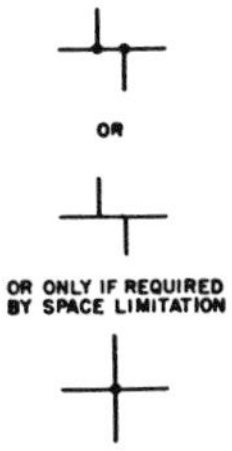

OR

OR ONLY IF REQUIRED
BY SPACE LIMITATION

43.7 Associated conductors

43.7.1 Pair (twisted unless otherwise specified)

P   OR   P

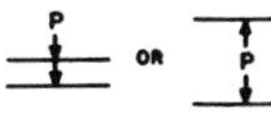

43.7.2 Triple (twisted unless otherwise specified)

T

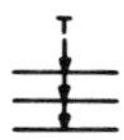

43.7.3 Quad

43.8 Assembled conductors; cable

Commonly used in communication diagrams.

43.8.1 Shielded single conductor

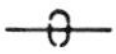

43.8.2 Coaxial cable
Coaxial transmission path

See note under item **43.1**.

43.8.3 2-conductor cable

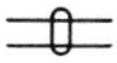

43.8.4 Shielded 2-conductor cable with shield grounded

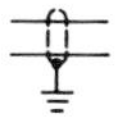

43.8.5 5-conductor cable

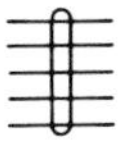

43.8.6 Shielded 5-conductor cable

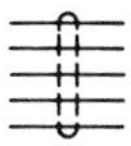

43.8.6.1 Shielded 5-conductor cable with conductors separated on the diagram for convenience

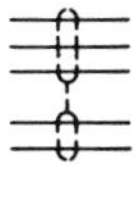

43.8.7 Cable underground or in conduit (*long dashes*)

Figure 2-12 (continued)

GRAPHICAL SYMBOLS
FOR ELECTRICAL DIAGRAMS
(in alphabetical order)

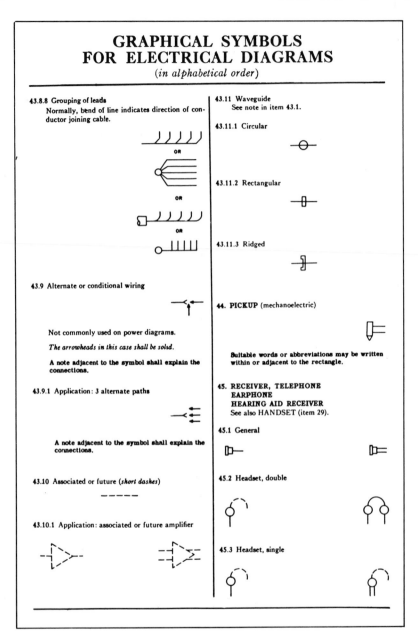

43.8.8 Grouping of leads
Normally, bend of line indicates direction of conductor joining cable.

OR

OR

OR

43.9 Alternate or conditional wiring

Not commonly used on power diagrams.

The arrowheads in this case shall be solid.

A note adjacent to the symbol shall explain the connections.

43.9.1 Application: 3 alternate paths

A note adjacent to the symbol shall explain the connections.

43.10 Associated or future (*short dashes*)

43.10.1 Application: associated or future amplifier

43.11 Waveguide
See note in item 43.1.

43.11.1 Circular

43.11.2 Rectangular

43.11.3 Ridged

44. PICKUP (mechanoelectric)

Suitable words or abbreviations may be written within or adjacent to the rectangle.

45. RECEIVER, TELEPHONE
EARPHONE
HEARING AID RECEIVER
See also HANDSET (item 29).

45.1 General

45.2 Headset, double

45.3 Headset, single

Figure 2-12 (continued)

GRAPHICAL SYMBOLS
FOR ELECTRICAL DIAGRAMS
(in alphabetical order)

46. RECTIFIER

46.1 Electron-tube rectifier

See TUBE, ELECTRON (item 64).

46.1.1 Pool-type-cathode power rectifier

46.2 Metallic rectifier; asymmetrical varistor; crystal diode; electrolytic rectifier

Arrow shows direction of forward (easy) current as indicated by direct-current ammeter.
The arrowhead in this case shall be filled.

46.2.1 Full-wave bridge type

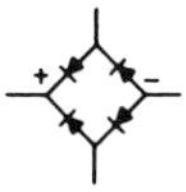

46.3 On connection or wiring diagrams, rectifier may be shown with terminals and polarity marking. Heavy line may be used to indicate nameplate or positive polarity end.

47. REGULATOR, SPEED (Contact-making governor)
Contacts open or closed as required; (shown here as closed).

48. RELAY

See also CONTACTOR (item 16)

Fundamental symbols for contacts, mechanical connections, coils, etc., are the basis of relay symbols and should be used to represent relays on complete diagrams.

The following letter combinations may be used with any relay symbol. The requisite number of these combinations may be used when a relay possesses more than one special feature.

AC	Alternating-current or ringing relay
D	Differential
DB	Double biased (biased in both directions)
DP	Dashpot
EP	Electrically polarized
†FO	Fast operate
†FR	Fast release
MG	Marginal
NB	No bias
NR	Nonreactive
P	Magnetically polarized using biasing spring, or having magnet bias
SA	Slow operate and slow release
SO	Slow operate
SR	Slow release
SW	Sandwich wound to improve balance to longitudinal currents

† Used where unusually fast operation or fast releasing is essential to the circuit operation.

The proper poling for a polarized relay shall be shown by the use of + and − designations applied to the winding leads. The interpretation of this shall be that current in the direction indicated shall move or tend to move the armature toward the contact shown nearest the core on the diagram. If the relay is equipped with numbered terminals, the proper terminal numbers shall also be shown.

48.1 Basic

Ⓡ

48.2 Relay coil

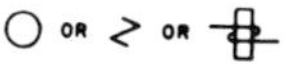

Always indicate the device designation within the circle.

48.2.1 Semicircular dot indicates inner end of winding

Figure 2-12 (continued)

GRAPHICAL SYMBOLS
FOR ELECTRICAL DIAGRAMS
(in alphabetical order)

48.3 Application: relay with transfer contacts

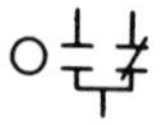

Always indicate the device designation within the circle.

OR

OR

48.4 Application: 2-pole double-make

48.5 Application: 1-pole double-break

48.6 Application: polarized relay with transfer contacts

48.7 Application: polarized (no bias) marginal relay with transfer contacts

48.8 Relay protective functions

The following symbols may be used to indicate protective functions, or device-function numbers (see latest edition of American Standard C37.2) may be placed in the circle or adjacent to the basic symbol.

48.8.1 Over, general

48.8.2 Under, general

48.8.3 Direction, general; directional over

48.8.4 Balance, general

48.8.5 Differential, general

48.8.6 Pilot wire, general

——— PW

48.8.7 Carrier current, general

——— CC

48.8.8 Operating quantity
The operating quantity is indicated by the following letters or symbols placed either on or above the center of the relay protective-function symbols shown above.

C *Current   GP Gas pressure S Synchronism
Z Distance   φ Phase   T Temperature
F Frequency W Power   V Voltage

* The use of the letter may be omitted in the case of current and the absence of such letter presupposes that the relay operates on current.

48.8.9 Ground relays
Relays operative on residual current only are so designated by attaching the ground symbol ⅠⅠ— to the relay protective-function symbol. Note that the zero phase-sequence designation given below may be used instead when desirable.

48.8.10 Phase sequence quantities
Operation on phase-sequence quantities may be indicated by the use of the conventional subscripts 0, 1, and 2 after the letter indicating the operating quantity.

48.8.11 Application

Figure 2-12 (continued)

GRAPHICAL SYMBOLS
FOR ELECTRICAL DIAGRAMS
(in alphabetical order)

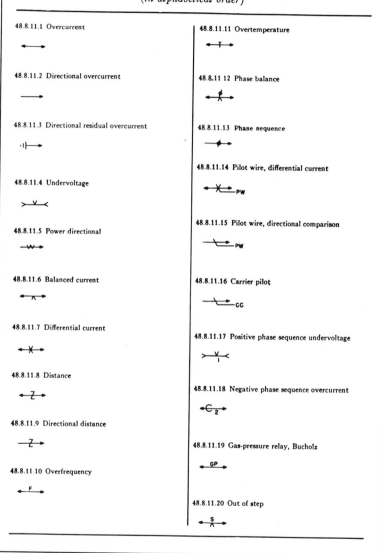

48.8.11.1 Overcurrent

48.8.11.2 Directional overcurrent

48.8.11.3 Directional residual overcurrent

48.8.11.4 Undervoltage

48.8.11.5 Power directional

48.8.11.6 Balanced current

48.8.11.7 Differential current

48.8.11.8 Distance

48.8.11.9 Directional distance

48.8.11.10 Overfrequency

48.8.11.11 Overtemperature

48.8.11.12 Phase balance

48.8.11.13 Phase sequence

48.8.11.14 Pilot wire, differential current

48.8.11.15 Pilot wire, directional comparison

48.8.11.16 Carrier pilot

48.8.11.17 Positive phase sequence undervoltage

48.8.11.18 Negative phase sequence overcurrent

48.8.11.19 Gas-pressure relay, Bucholz

48.8.11.20 Out of step

Figure 2-12 (continued)

GRAPHICAL SYMBOLS
FOR ELECTRICAL DIAGRAMS
(in alphabetical order)

49. REPEATER

49.1 1-way repeater

Triangle points in the direction of transmission.

49.2 2-wire 2-way repeater

49.2.1 2-wire 2-way repeater with low-frequency bypass

49.3 4-wire 2-way repeater

50. RESISTOR
See also TERMINATION (item 59).

For resistors with nonlinear characteristics, see BALLAST LAMP (item 33.1), THERMISTOR (item 60), and VARISTOR (item 66).

Always add identification within or adjacent to the rectangle.

50.1 General

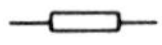

For Electronics Application

50.2 Tapped resistor

For Electronics Application

50.3 Application: with adjustable contact

For Electronics Application

50.4 Application: adjustable or continuously adjustable (variable) resistor

For Electronics Application

50.5 Heating resistor

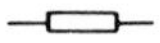

For Electronics Application

50.6 Instrument or relay shunt

Connect instrument or relay to terminals in the box.

50.7 Shunt resistor

For Electronics Application

Figure 2-12 (continued)

GRAPHICAL SYMBOLS
FOR ELECTRICAL DIAGRAMS
(in alphabetical order)

51. RESONATOR
Excluding piezoelectric and magnetostriction devices.

51.1 General

Commonly used for coaxial and waveguide transmission.

51.2 Applications

51.2.1 Resonator with mode suppression coupled by an *E*-plane aperture to a guided transmission path and by a loop to a coaxial path.

51.2.2 Tunable resonator having adjustable *Q* coupled by a probe to a coaxial system.

51.2.3 Tunable resonator with direct-current ground connected to an electron device and adjustably coupled by an *E*-plane aperture to a rectangular waveguide.

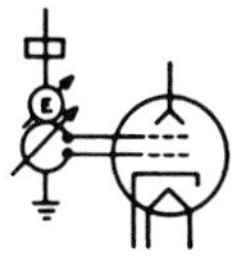

52. SHIELD
SHIELDING *(short dashes)*
Normally used for electric or magnetic shielding. When used for other shielding, a note should so indicate. For typical applications see:
CAPACITOR (item 8.1.1)
PATH, TRANSMISSION (items 43.8.1, 43.8.4, and 43.8.6)
TRANSFORMER (items 63.2.1 and 63.2.2)
TUBE, ELECTRON (item 64.7)

53. SHIFTER, PHASE
For power circuits see MACHINE, ROTATING (items 35.10.17 and 35.10.18).

53.1 General

53.2 3-wire or 3-phase

53.2.1 Application: adjustable

54. SOUNDER, TELEGRAPH

55. SUPPRESSION, MODE
Commonly used in coaxial and waveguide transmission.

56. SWITCH
See also FUSE (item 27); CONTACT, ELECTRIC (item 15);

Fundamental symbols for contacts, mechanical connections, etc., may be used for switch symbols.

The standard method of showing switches is in a position with no operating force applied. For switches that may be in any one of two or more positions with no operating force applied and for switches actuated by some mechanical device (as in air-pressure, liquid-level, rate-of-flow, etc., switches), a clarifying note may be necessary to explain the point at which the switch functions.

When the basic switch symbols in items 56.1 through 56.4 are shown on a diagram in the closed position, terminals must be added for clarity.

Figure 2-12 (continued)

GRAPHICAL SYMBOLS
FOR ELECTRICAL DIAGRAMS
(in alphabetical order)

56.1 Single throw, general

56.2 Double throw, general

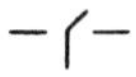

56.2.1 Application   2-pole double-throw switch with
terminals shown

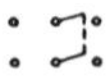

56.3 Knife switch, general

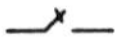

56.3.1 Application. 3-pole double-throw knife switch
with auxiliary contacts and terminals

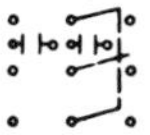

56.3.2 Application: 2-pole field-discharge knife switch
with terminals and discharge resistor

Always add identification within or adjacent to the
rectangle.

56.4 Switch with horn gap

56.5 Sector switch

56.6 Push button, momentary or spring return

56.6.1 Circuit closing (make)

56.6.2 Circuit opening (break)

56.6.3 Two-circuit

56.7 Push button, maintained or not spring return

56.7.1 Two circuit

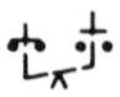

56.8 Switch, nonlocking; momentary or spring return

The symbols to the left are commonly used for
spring buildups in key switches, relays, and jacks.

The symbols to the right are commonly used for
toggle switches.

56.8.1 Circuit closing (make)

56.8.2 Circuit opening (break)

56.8.3 Two-circuit

56.8.4 Transfer

56.8.5 Make-before-break

Figure 2-12 (continued)

GRAPHICAL SYMBOLS
FOR ELECTRICAL DIAGRAMS
(in alphabetical order)

56.9 Switch, locking

The symbols to the left are commonly used for spring buildups in key switches, relays, and jacks.

The symbols to the right are commonly used for toggle switches.

56.9.1 Circuit closing (make)

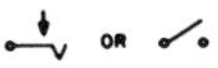

56.9.2 Circuit opening (break)

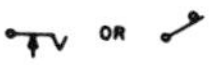

56.9.3 Transfer, 2-position

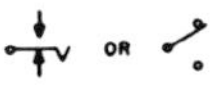

56.9.4 Transfer, 3-position

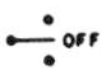

56.9.5 Make-before-break

56.10 Switch, combination locking and nonlocking
See also item 56.11.
Commonly used for toggle-switches.

56.10.1 3-position 1-pole circuit closing (make), off, momentary circuit closing (make)

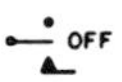

56.10.2 3-position 2 pole: circuit closing (make), off, momentary circuit closing (make)

56.11 Switch, key-type, applications

56.11.1 2-position with locking transfer and break contacts

56.11.2 3-position with nonlocking transfer and locking break contacts

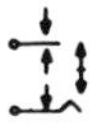

56.11.3 3-position, multicontact combination

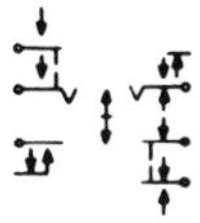

56.11.4 2-position, half of key switch normally operated, multicontact combination

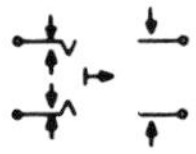

56.12 Selector or multiposition switch

The position in which the switch is shown may be indicated by a note or designation of switch position.

56.12.1 General (for power and control diagrams)
Any number of transmission paths may be shown.

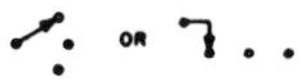

56.12.2 Break-before-make, nonshorting (nonbridging) during contact transfer

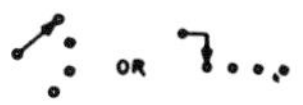

Figure 2-12 (continued)

GRAPHICAL SYMBOLS
FOR ELECTRICAL DIAGRAMS
(in alphabetical order)

56.12.3 Make-before-break, shorting (bridging) during contact transfer

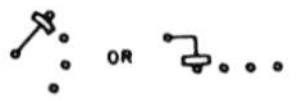

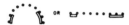

56.12.4 Segmental contact

56.12.5 22-point selector switch

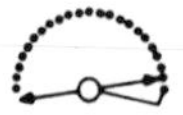

56.12.6 10-point selector switch with fixed segment

56.12.7 Wafer, 3-pole 3-circuit with 2 nonshorting and 1 shorting moving contacts
Viewed from end opposite control knob or actuator unless otherwise indicated.
For more than one section, section No. 1 is nearest control knob.

When contacts are on both sides, front contacts are nearest control knob.

56.12.8 Slide switch, typical ladder-type interlock
In the example, one slide is shown operated.
Slides are shown in released position unless otherwise noted.

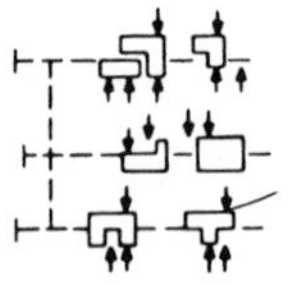

56.12.9 Master or control switch

A table of contact operation must be shown on the diagram. A typical table is shown below.

DETACHED CONTACTS
SHOWN ELSEWHERE
ON DIAGRAM

CONTACT	POSITION		
	A	B	C
1-2			X
3-4	X		
5-6			X
7-8	X		

X INDICATES CONTACT CLOSED

HANDLE END

FOR CONNECTION
OR WIRING DIAGRAM

56.12.10 Master or control switch
(Cam-operated contact assembly) 6-circuit 3-point reversing switch.
A table of contact operation must be shown on the diagram. A typical table is shown below.
Tabulate special features in note.

DETACHED CONTACTS
SHOWN ELSEWHERE
ON DIAGRAM

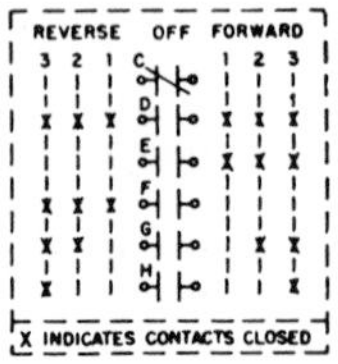

X INDICATES CONTACTS CLOSED

HANDLE END

FOR CONNECTION
OR WIRING DIAGRAM

Figure 2-12 (continued)

GRAPHICAL SYMBOLS
FOR ELECTRICAL DIAGRAMS
(in alphabetical order)

56.12.11 Drum switch, sliding-contact type, typical example

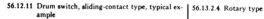
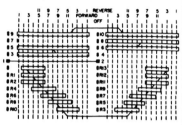

56.13 Switches with specific features

56.13.1 Key-operated lock switch
Use appropriate standard symbol and add key designation or other information in note.

56.13.2 Limit switch

56.13.2.1 General
Use appropriate standard symbol and identify by LS or other suitable note.

56.13.2.2 Track-type; circuit-opening contact

∅Identify by LS or other suitable note.

56.13.2.3 Lead-screw type; circuit-opening contacts

∅Identify by LS or other suitable note.

56.13.2.4 Rotary type

∅Identify by LS or other suitable note.

56.13.3 Mushroom-head safety feature
Application to 2-circuit push-button switch.

56.13.4 Safety interlock

56.13.4.1 General
If specific type identification is not required, use applicable standard symbol.

56.13.4.2 If specific type identification is required; circuit opening

56.13.4.3 If specific type identification is required; circuit closing

56.13.5 Hook switch

56.13.6 Dial switch, telephone type

TYPICAL

56.13.7 Switch in evacuated envelope, 1-pole double-throw

Figure 2-12 (continued)

GRAPHICAL SYMBOLS
FOR ELECTRICAL DIAGRAMS

(in alphabetical order)

57. SYNCHRO
SYNCHRO CONTROL TRANSFORMER
SYNCHRO RECEIVER
SYNCHRO TRANSMITTER

If identification is required, a letter combination from the following list shall be placed adjacent to the symbol to indicate the type of synchro.

CDX   Control-differential synchro transmitter
CT   Synchro control transformer
CX   Synchro control transmitter
TDR   Torque-differential synchro receiver
TDX   Torque-differential synchro transmitter
TR   Torque-synchro receiver
TX   Torque-synchro transmitter

If the outer winding is rotatable in bearings, the suffix B shall be added to the above letter combinations.

57.1 Synchro control transformer
Synchro receiver
Synchro transmitter

57.2 Differential synchro receiver
Differential synchro transmitter

58. TERMINAL, CIRCUIT
See also TUBE TERMINALS (item 64.12.2).

58.1 Terminal board or terminal strip with 4 terminals shown; group of 4 terminals

Number and arrangement as convenient.

59. TERMINATION

59.1 Cable termination

Line on left of symbol shown indicates cable.

59.2 Open circuit (open)

Not a fault.

Commonly used in coaxial and waveguide diagrams.

59.3 Short circuit (short)

Not a fault.

Commonly used in coaxial and waveguide diagrams.

59.3.1 Application: movable short

59.4 Terminating capacitor

Commonly used in coaxial and waveguide diagrams.

59.4.1 Application: series capacitor and path open

59.4.2 Application: series capacitor and path shorted

59.5 Terminating inductor

Commonly used in coaxial and waveguide diagrams.

59.5.1 Application: series inductor and path open

59.5.2 Application: series inductor and path shorted

Figure 2-12 (continued)

GRAPHICAL SYMBOLS
FOR ELECTRICAL DIAGRAMS
(in alphabetical order)

59.6 Terminating resistor

Commonly used in coaxial and waveguide diagrams.

59.6.1 Application: series resistor and path open

59.6.2 Application: series resistor and path shorted

60. THERMISTOR
T indicates that the primary characteristic of the element within the circle is designed to vary with temperature.

60.1 General

60.2 With independent integral heater

61. THERMOCOUPLE

61.1 Dissimilar-metals device

61.1.1 Temperature-measuring thermocouple

61.1.2 Current-measuring thermocouple
Explanatory words and arrows are not a part of the symbols shown.

61.1.2.1 Thermocouple with integral heater internally connected

61.1.2.2 Thermocouple with integral insulated heater

61.2 Semiconductor device

61.2.1 Temperature-measuring semiconductor thermocouple

61.2.2 Current-measuring semiconductor thermocouple

62. TRANSDUCER, MODE
Commonly used in coaxial and waveguide diagrams.

62.1 General

62.2 Application: transducer from rectangular to circular waveguide

62.3 Application: transducer from rectangular waveguide to coaxial with mode suppression and direct-current grounds connected

Figure 2-12 (continued)

GRAPHICAL SYMBOLS
FOR ELECTRICAL DIAGRAMS
(in alphabetical order)

63. TRANSFORMER

63.1 General

Additional windings may be shown or indicated by a note.

For power transformers, use polarity marking H_1-X_1, etc., from American Standard C6.1. For polarity markings on current and potential transformers, see items 63.16.1 and 63.17.1.

In coaxial and waveguide circuits, this symbol will represent a taper or step transformer without mode change.

63.1.1 Application: transformer with direct-current ground connections and mode suppression between two rectangular waveguides

63.2 If it is desired especially to distinguish a magnetic-core transformer

63.2.1 Application: shielded transformer with magnetic core shown

63.2.2 Application: transformer with magnetic core shown and with a shield between windings. The shield is shown connected to the frame.

63.3 One winding with adjustable inductance

63.4 Each winding with separately adjustable inductance

63.5 Adjustable mutual inductor, constant-current transformer

63.6 With taps, 1-phase

63.7 Autotransformer, 1-phase

63.7.1 Adjustable

63.8 Step-voltage regulator or load-ratio control autotransformer

63.9 Load-ratio control transformer with taps

Figure 2-12 (continued)

GRAPHICAL SYMBOLS
FOR ELECTRICAL DIAGRAMS
(in alphabetical order)

63.10 1-phase induction voltage regulator(s)

Number of regulators may be written adjacent to the symbol.

63.11 Triplex induction voltage regulator

63.12 3-phase induction voltage regulator

63.13 1-phase 2-winding transformer

63.13.1 3-phase bank of 1-phase 2-winding transformers See latest edition of American Standard C6.1 for interconnection conventions for complete symbols.

63.14 Polyphase transformer

OR

63.15 1-phase 3-winding transformer

OR   OR

63.16 Current transformer(s)

OR   OR

63.16.1 Current transformer with polarity marking. Instantaneous direction of current into one polarity mark corresponds to current out of the other polarity mark.

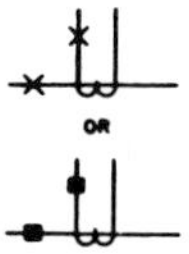

OR

Figure 2-12 (continued)

GRAPHICAL SYMBOLS
FOR ELECTRICAL DIAGRAMS
(in alphabetical order)

63.16.2 Bushing-type current transformer

The broken line - — - indicates where line connection to a symbol is made and is not a part of the
symbol.

63.17 Potential transformer(s)

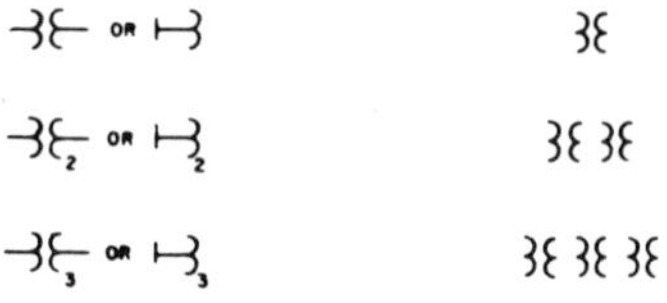

63.17.1 Potential transformer with polarity mark. Instantaneous direction of current into one polarity mark corresponds to current out of the
other polarity mark.

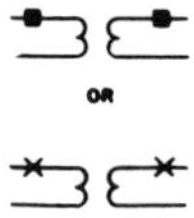

OR

63.18 Outdoor metering device

SHOW ACTUAL
CONNECTION
INSIDE BORDER

63.19 Transformer winding connection symbols
For use adjacent to the symbols for the transformer windings.

63.19.1 2-phase 3-wire, ungrounded

L

63.19.1.1 2-phase 3-wire, grounded

63.19.2 2-phase 4-wire

63.19.2.1 2-phase 5-wire, grounded

63.19.3 3-phase 3-wire, delta or mesh

△

63.19.3.1 3-phase 3-wire, delta, grounded

63.19.4 3-phase 4-wire, delta, ungrounded

63.19.4.1 3-phase 4-wire, delta, grounded

63.19.5 3-phase, open-delta

∠

63.19.5.1 3-phase, open-delta, grounded at common
point

63.19.5.2 3-phase, open-delta, grounded at middle point
of one transformer

63.19.6 3-phase, broken-delta

△

63.19.7 3-phase, wye or star, ungrounded

⅄

Figure 2-12 (continued)

GRAPHICAL SYMBOLS
FOR ELECTRICAL DIAGRAMS

(in alphabetical order)

63.19.7.1 3-phase, wye, grounded neutral
The direction of the stroke representing the neutral can be arbitrarily chosen.

63.19.8 3-phase 4-wire, ungrounded

63.19.9 3-phase, zigzag, ungrounded

63.19.9.1 3-phase, zigzag, grounded

63.19.10 3-phase, Scott or T

63.19.11 6-phase, double-delta

63.19.12 6-phase, hexagonal (or chordal)

63.19.13 6-phase, star (or diametrical)

63.19.13.1 6-phase, star, with grounded neutral

64. TUBE, ELECTRON
Tube-component symbols are shown first. These are followed by typical applications showing the use of these specific symbols in the various classes of devices such as thermionic, cold-cathode, and photoemissive tubes of varying structures and combinations of elements (triodes, pentodes, cathode-ray tubes, magnetrons, etc.).

Lines outside of the envelope are not part of the symbol but are electrical connections thereto.

Connections between the external circuit and electron tube symbols within the envelope may be located as required to simplify the diagram.

64.1 Emitting electrode

64.1.1 Directly heated (filamentary) cathode
Note—Leads may be connected in any convenient manner to ends of the ∧ provided the identity of the ∧ is retained.

64.1.1.1 With tap
See note in item **64.10.3**.

64.1.2 Indirectly heated cathode
Lead may be connected to either extreme end of the ⌐ or, if required, to both ends, in any convenient manner.

64.1.3 Cold cathode (including ionically heated cathode)

64.1.4 Photocathode

64.1.5 Pool cathode

64.1.6 Ionically heated cathode with provision for supplementary heating
See note in item **64.1.1**

Figure 2-12 (continued)

GRAPHICAL SYMBOLS
FOR ELECTRICAL DIAGRAMS
(in alphabetical order)

64.2 Controlling electrode

64.2.1 Grid (including beam-confining or beam-forming electrodes)

64.2.2 Deflecting electrodes (used in pairs); reflecting or repelling electrode (used in velocity-modulated tube)

64.2.3 Ignitor (in pool tubes) (should extend into pool) Starter (in gas tubes)

64.2.4 Excitor (contactor type)

64.3 Collecting electrode

64.3.1 Anode or plate (including collecting electrode and fluorescent target)

64.3.2 Target or X-ray anode

Drawn at about a 45-degree angle.

64.4 Collecting and emitting electrode

64.4.1 Dynode

64.4.2 Alternately collecting and emitting

64.4.2.1 Composite anode-photocathode

64.4.2.2 Composite anode-cold cathode

64.4.2.3 Composite anode-ionically heated cathode with provision for supplementary heating

See note in item 64.1.1

64.5 Heater
See note in item 64.1.1

64.5.1 With tap
See item 64.10.3.

64.6 Envelope (shell)

The general envelope symbol identifies the envelope or enclosure regardless of evacuation or pressure. When used with electron-tube component symbols, the general envelope symbol indicates a vacuum enclosure unless otherwise specified. A gas-filled electron device may be indicated by a dot within the envelope symbol.

64.6.1 General

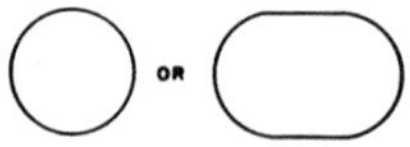

OR

64.6.1.1 Split envelope

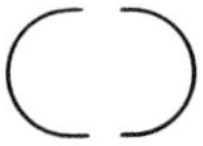

If necessary, envelope may be split.

64.6.2 Gas-filled

The dot may be located as convenient.

Figure 2-12 (continued)

GRAPHICAL SYMBOLS
FOR ELECTRICAL DIAGRAMS

(in alphabetical order)

64.7 Shield
See item 64.10.10.

This is understood to shield against electric fields unless otherwise noted.

64.7.1 Any shield against electric fields that is within the envelope and that is connected to an independent terminal

64.7.2 Outside envelope of X-ray tube

64.8 Coupling
See COUPLING (item 20) and PATH, TRANSMISSION (items 43.8.2 and 43.11).

64.8.1 Coupling by loop (electromagnetic type)

Coupling loop may be shown inside or outside envelope as desired, but if inside it should be shown grounded.

64.9 Resonators (cavity type)

64.9.1 Single-cavity envelope and grid-type associated electrodes

64.9.2 Double-cavity envelope and grid-type associated electrodes

64.9.3 Multicavity magnetron anode and envelope

64.10 General notes

64.10.1 If new symbols are necessary, they should be formed where possible from component symbols. For example, see DYNODE (item 64.4.1), which combines the anode and photocathode convention.

64.10.2 A connection to anode, dynode, pool cathode, photocathode, deflecting electrode, composite anode-photocathode, and composite anode-cold cathode shall be to the center of that symbol. Connection to any other electrode may be shown at either end or both ends of the electrode symbol.

64.10.3 A diagram for a tube having more than one heater or filament shall show only one heater or filament symbol ⋀ unless they have entirely separate connections. If a heater or filament tap is made, either brought out to a terminal or internally connected to another element, it shall be connected at the vertex of the symbol, regardless of the actual division of voltage across the heater or filament.

64.10.4 Standard symbols, such as the inclined arrow for tunability and connecting dotted lines for ganged components, may be added to a tube symbol to extend the meaning of the tube symbol, provided such added feature or component is integral with the tube.

64.10.5 Electric components, such as resistors, capacitors, or inductors, which are integral parts of the tube and are important to its functional operation, shall be shown in the standard manner.

64.10.6 Multiple equipotential cathodes that are directly connected inside the tube shall be shown as a single cathode.

64.10.7 A tube having two or more grids tied internally shall be shown with symbols for each grid, except when the grids are adjacent in the tube structure. Thus, the diagram for a twin pentode having a common screen-grid connection for each section and for a converter tube having the No. 3 and No. 5 grids connected internally will show separate symbols for each grid. However, a triode where the control grid is physically in the form of two grid windings would show only one grid.

64.10.8 A tube having a grid adjacent to a plate but internally connected to the plate to form a portion of it shall be shown as having a plate only.

Figure 2-12 (continued)

GRAPHICAL SYMBOLS
FOR ELECTRICAL DIAGRAMS
(in alphabetical order)

64.10.9 Associated parts of a circuit, such as focusing coils, deflecting coils, field coils, etc., are not a part of the tube symbol but may be added to the circuit in the form of standard symbols. For example, resonant-type magnetron with permanent magnet may be shown:

64.10.10 External and internal shields, whether integral parts of tubes or not, shall be omitted from the circuit diagram unless the circuit diagram requires their inclusion.

64.10.11 In line with standard drafting practice, straight-line crossovers are recommended.

64.11 Typical applications

64.11.1 Triode with directly heated filamentary cathode and envelope connection to base terminal

64.11.2 Equipotential-cathode pentode showing use of elongated envelope

64.11.3 Equipotential-cathode twin triode illustrating elongated envelope and rule of item 64.10.3.

64.11.4 Typical wiring figure
This figure illustrates how tube symbols may be placed in any convenient position in a circuit.

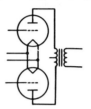

64.11.5 Cold-cathode gas-filled tube

64.11.5.1 Rectifier; voltage regulator for direct-current operation
See also GLOW LAMP (item 33.3).

64.11.6 Phototube

64.11.6.1 Single-unit, vacuum type

64.11.6.2 Multiplier type

64.11.7 Cathode-ray tube

64.11.7.1 With electric-field deflection

64.11.7.2 For magnetic deflection

Figure 2-12 (continued)

GRAPHICAL SYMBOLS
FOR ELECTRICAL DIAGRAMS
(in alphabetical order)

64.11.8 Mercury-pool tube
See also RECTIFIER (item 46.1.1).

64.11.8.1 With ignitor and control grid

64.11.8.2 With excitor, control grid, and holding anode

64.11.8.3 Single-anode pool-type vapor rectifier with ignitor

64.11.8.4 6-anode metallic-tank pool-type vapor rectifier with excitor, showing rigid-terminal symbol for control connection to tank (pool cathode is insulated from tank)

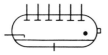

Anode symbols are located as convenient.

64.11.9 Magnetron

64.11.9.1 Resonant type with coaxial output

64.11.9.2 Transit-time split-plate type with stabilizing deflecting electrodes and internal circuit

64.11.9.3 Tunable, aperture coupled

64.11.10 Velocity-modulation (velocity-variation) tube

64.11.10.1 Reflex klystron, integral cavity, aperture coupled

64.11.10.2 Double-cavity klystron, integral cavity, permanent external-ganged tuning, loop coupled (coupling loop may be shown inside if desired. See item 64.8.1)

64.11.11 Transmit-receive (TR) tube
Gas filled, tunable integral cavity, aperture coupled, with starter.

Figure 2-12 (continued)

GRAPHICAL SYMBOLS
FOR ELECTRICAL DIAGRAMS
(in alphabetical order)

64.11.12 X-ray tube

64.11.12.1 With filamentary cathode and focusing grid (cup). The anode may be cooled by fluid or radiation.

64.11.12.2 With control grid, filamentary cathode, and focusing cup

64.11.12.3 With grounded electrostatic shield

64.11.12.4 Double focus with rotating anode (see note in item 64.10.9)

64.11.12.5 With multiple accelerating electrode, electrostatically and electromagnetically focused (see note in item 64.10.9)

64.12 Basing and terminal connections for connection (wiring) diagrams

Not normally used for schematic diagrams.

64.12.1 Basing orientation symbols

64.12.1.1 For tubes with keyed bases
Explanatory word and arrow are not a part of the symbol shown.

64.12.1.2 For tubes with bayonets, bosses, and other reference points

64.12.2 Tube terminals
The usage of the rigid-envelope-terminal symbol of item 64.12.2.2 includes the indication of any external metallic envelope or conducting coating or casing that has a contact area (as in cathode-ray tubes, metallic "pencil" tubes, etc.). However, where contact to such external metallic elements is made through a base terminal, a dot junction is employed as in item 64.12.3.1 to indicate that voltage applied to this base terminal may make the envelope alive.

Terminal symbols may be added to the composite device symbols where desired without changing the meaning or becoming a part of the symbol.

64.12.2.1 Base terminals
Explanatory words and arrows are not a part of the symbol.

Figure 2-12 (continued)

GRAPHICAL SYMBOLS
FOR ELECTRICAL DIAGRAMS
(in alphabetical order)

64.12.2.2 Envelope terminals
Explanatory words and arrows are not a part of the symbol.

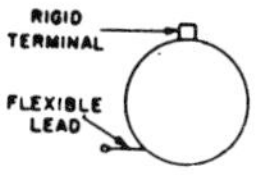

64.12.3 Applications

64.12.3.1 Triode with indirectly heated cathode and envelope connected to base terminal

64.12.3.2 Triode-heptode with rigid envelope connection

64.12.3.3 Ultra-high-frequency triode (disk-seal-tube type) with internal capacitor

64.12.3.4 Rectifier with heater tap and envelope connected to base terminal

64.12.3.5 Equipotential-cathode twin triode with tapped heater

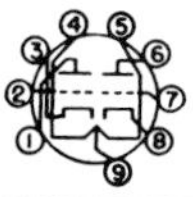

65. UNIT, PIEZOELECTRIC CRYSTAL

66. VARISTOR
See also RECTIFIER (item 46)

Electroelectrical transducer with nonlinear characteristics.

The arrowheads in these cases are to be filled.

66.1 Asymmetrical; metallic rectifier

Arrow shows direction of forward (easy) current as indicated by direct-current ammeter.

66.2 Symmetrical

67. VIBRATOR

(Not applicable to Military drawings; Refer to MIL STD-15A)

67.1 Typical shunt drive (contacts as required) (with terminals shown)

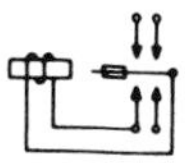

67.2 Typical separate drive (contacts as required) (with terminals shown)

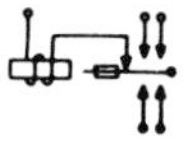

Figure 2-12 (continued)

The most common abbreviations found in control diagrams include the following:

1. *SP:* single pole
2. *ST:* single throw
3. *DP:* double pole
4. *DT:* double throw
5. *3P:* three pole
6. *2P:* two pole
7. *NC:* normally closed contact
8. *NO:* normally open contact

Normally closed contact means that the contact is closed when the relay coil is not energized. *Normally open contact* means that the contact is open when the relay coil is not energized. Contacts will change position when the relay coil is energized. The normally closed contacts will open and the normally open contacts will close.

Some examples of control symbols and abbreviations are shown in the illustrations to follow. For example, the drawing in Fig. 2-13 represents a single-pole single-throw (SPST) switch; the contact on the left is normally open and the one on

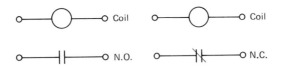

Figure 2-13   Single-pole single-throw switch.

the right is normally closed. The circuit in Fig. 2-14 represents a single-pole double-throw (SPDT) switch and the one in Fig. 2-15 shows double-pole single-throw (DPST) contacts—one group of the normally open type and the other of the normally closed type. A double-pole double-throw (DPDT) circuit is shown in Fig. 2-16 and a three-pole double-throw (3PDT) circuit is represented in Fig. 2-17.

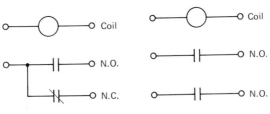

Figure 2-14   Single-pole double-throw switch.

Figure 2-15   Double-pole single-throw contacts.

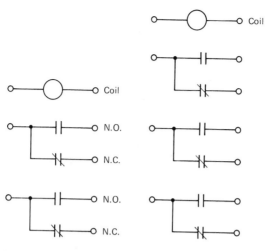

Figure 2-16 Double-pole double-throw circuit.

Figure 2-17 Three-pole double-throw circuit.

SCHEMATIC WIRING DIAGRAMS

Schematic wiring diagrams represent components in the control system by symbols, the wiring and connection to each component, and other detailed information. Sometimes the conductors are shown in an assembly of several wires which appear as one line on the drawing. When this method is used, each wire should be numbered where it enters the assembly and should keep the same number when it comes out of the assembly to be connected to a component in the system. When reading or using such drawings, if the schematic does not follow this procedure, mark and number the wires yourself.

Although the symbols represent certain components, an exact description of each is usually listed in schedules or noted on the drawings. Such drawings are seldom, if ever, drawn to scale as an architectural or cabinet drawing would be. They appear in diagrammatic form. In better drawings, however, the components are arranged in a neat and logical sequence so that they are easily traced and can be understood easily.

Electronic schematic diagrams indicate the scheme or plan according to which electronic or control components are connected for a specific purpose. Diagrams are not normally drawn to scale, and the symbols rarely look exactly like the component. Lines joining the symbols representing electronic or control components indicate that the components are connected.

To serve all its intended purposes, a schematic diagram must be accurate. Also, it must be understood by all qualified personnel, and it must provide definite information without ambiguity.

The schematics for a control circuit should indicate all circuits in the device. If they are accurate and well prepared, it will be easy to read and follow an entire closed path in each circuit. If there are interconnections, they will be clearly indicated. In almost all cases the conductors connecting the electronic symbols will be drawn either horizontally or vertically. Rarely are they slanted. A dot at the junction of two crossing wires means a connection between the two wires. An absence of a dot usually indicates that the wires cross without connecting.

Schematic diagrams are, in effect, shorthand explanations of the manner in which an electronic circuit or group of circuits operates. They make extensive use of symbols and abbreviations. The more commonly used symbols were explained earlier in this chapter. These symbols must be learned if one is to interpret control drawings with the necessary speed required in the field or design department. The use of symbols presumes that the person reading the diagram is reasonably familiar with the operation of the device and that he or she will be able to assign the correct meaning to the symbols. If the symbols are unusual, a legend will normally be provided to clarify matters.

Every component on a complete schematic diagram usually has a number to identify the component. Supplementary data about such parts are supplied on the diagram or on an accompanying list in the form of a schedule, which describes the component in detail or refers to a common catalog number familiar in the trade.

To interpret schematic diagrams, remember that each circuit must be complete in itself. Each component should be in a closed loop connected by conductors to a source of electric current such as a transformer or line voltage. There will

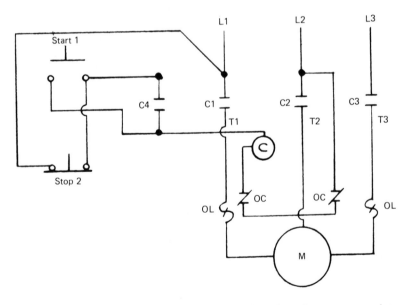

Figure 2-18 Complete schematic wiring diagram for a three-phase ac nonreversing motor starter.

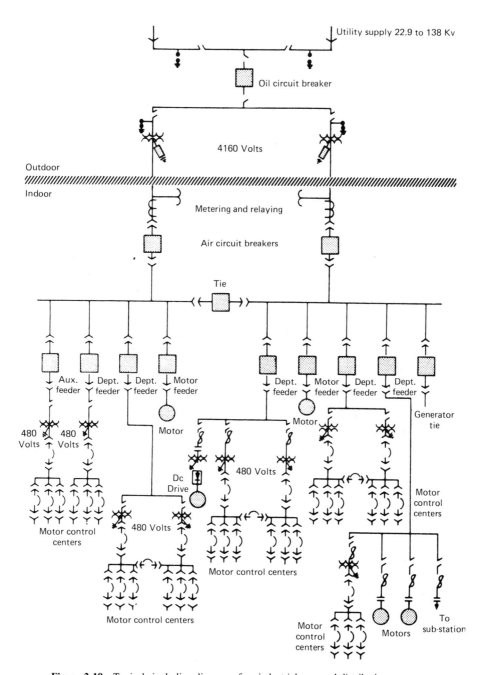

Figure 2-19 Typical single-line diagram of an industrial-powered distribution system.

always be a conducting path leading from the source to the component and a return path leading from the component to the source. The path may consist of one or more conductors. Other components may also be in the same loop or in additional loops branching off to other devices. For each electronic component, it must be possible to trace a completed conducting loop to the source.

ELECTRICAL WIRING DIAGRAMS

Complete schematic electrical wiring diagrams used in highly complex motor control circuits are also represented by symbols. Every wire is either shown by itself or included in an assembly of several wires that appear as one line on the drawing. Figure 2-18 shows a complete schematic wiring diagram for a three-phase ac magnetic nonreversing motor starter.

Note that this diagram shows the various devices in symbol form and indicates the actual connections of all wires between the devices. The three-wire supply lines are indicated by L_1, L_2, and L_3. The motor terminals of motor M are indicated by T_1, T_2, and T_3. Each line has a thermal overload-protection device (OL) connected in series with normally open line contactors C_1, C_2, and C_3, which are controlled by the magnetic starter coil, C. Each contactor has a pair of contacts that closes or opens during operation. The control station, consisting of start pushbutton 1 and stop pushbutton 2, is connected across lines L_1 and L_2. An auxiliary contactor C_4 is connected in series with the stop pushbutton and in parallel with the start pushbutton. The control circuit also has normally closed overload contactors (OC) connected in series with the magnetic starter coil (C).

Figure 2-19 shows a typical single-line diagram of an industrial power distribution system. In analyzing this diagram, the utility company will bring its lines to a substation outside the plant building. Air switches, lightning arresters, single-throw switches, and an oil circuit breaker are provided there. This substation also reduces the primary voltage to 4160 V by transformers. Again, lightning arresters and various other disconnecting means are shown.

CONCLUSION

Reading electrical/electronic drawings may seem difficult to the beginner, but after some practice, the reading of such drawings soon becomes second nature, and these drawings will be read as one would read a conventional set of instructions in one's native language.

chapter three

Electricity Basics

Many different electrical power supplies will be encountered while working with motor controls. The type supplied is dependent on either the power company's distribution network and transformer hookup or on the manufacturing plant's generation system.

Electric power is normally distributed at high voltages for reasons of economics; that is, the higher the voltage, the smaller the wire size necessary to carry the same amount of power. The power company uses transformers at generators to step up the voltage for cross-country distribution, and transformers are used again at substations or at the point of utilization to step the voltage down to the amount required. Most power is generated as three-phase with one of the phases used to obtain single-phase.

From a practical standpoint, people involved with motor controls need only be concerned with the power supply on the secondary (usage) side of the transformer, for this determines the characteristic of the power supply for use in the building or on the premises.

There are two general arrangements of transformers and secondaries in common use. The first arrangement is the sectional form, in which a unit of load, such as one city street or city block, is served by a fixed length of secondary, with the transformer located in the middle. The second arrangement is the continuous form, where the secondary is installed in one long continuous run—with transformers spaced along it at the most suitable points. As the load grows or shifts, the transformers can be moved or rearranged, if desired. In sectional arrangement, such a load can be cared for only by changing to a larger transformer or by installing an additional unit in the same section.

One of the greatest advantages of the secondary bank is that the starting currents of motors are divided between transformers, reducing the voltage drop and diminishing the resulting lamp flicker at the various outlets.

Power companies in the United States and Canada are now incorporating networks into their secondary power systems, especially in areas where a high degree of service reliability is necessary. Around cities and industrial applications, most secondary circuits are three-phase—either 120/208 V or 480/208 V, wye-connected. Usually, two to four primary feeders are run into the area, and transformers are connected alternately to them. The feeders are interconnected in a grid, or network, so that if any feeder were to go out of service, the load would still be carried by the remaining feeders.

The primary feeders supplying networks are run from substations at the usual primary voltage for the system, such as 4160, 4800, 6900, or 13,200 V. Higher voltages are practicable if the loads are large enough to warrant them.

COMMON POWER SUPPLIES

The most common power supply used for residential and small commercial applications is the 120/240-V single-phase service used primarily for light and power, including single-phase motors up to about $7\frac{1}{2}$ hp. A diagram of this service is shown in Fig. 3-1.

Four-wire delta-connected secondaries (Fig. 3-2) and four-wire wye-connected secondaries (Fig. 3-3) are common around industrial and large commercial applications.

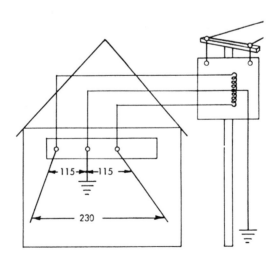

Figure 3-1   Single-phase three-wire. (Courtesy Borg-Warner Air Conditioning.)

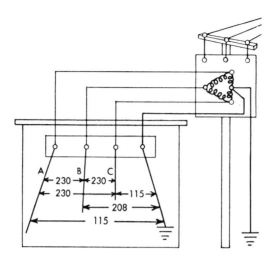

Figure 3-2   Four-wire delta (grounded). (Courtesy Borg-Warner Air Conditioning.)

TRANSFORMERS

The main purpose of a power transformer is to obtain a voltage supply that is different from the main voltage available. A transformer's capacity is rated in kVA (kilovolt-amperes). A transformer must be designed for the frequency (in hertz) of the system involved.

To obtain a certain voltage when the available supply is other than the voltage

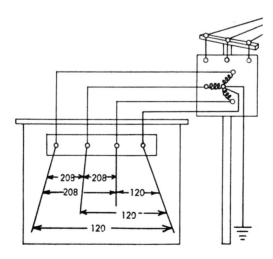

Figure 3-3   Four-wire WYE (grounded). (Courtesy Borg-Warner Air Conditioning.)

required, either a step-up or a step-down transformer may be used, depending on the circumstance. Each type is interchangeable; that is, primary and secondary windings may be reversed.

When selecting transformers, the supply line voltage, load voltage requirement, and the load amperage requirement must be determined. Then manufacturers' catalogs and/or manufacturers' representatives must be consulted for a specific selection.

At times it may be necessary to obtain a relatively small voltage correction when dealing with motors and motor controls. For example, a motor rated for 240 V may be connected to a 208-V power supply. To obtain this additional voltage, a buck-and-boost transformer may be used. In this case, it will be used to boost the voltage. If the reverse were true (a 208-V motor used on a 240-V supply), the transformer would be used to "buck" the voltage (lower it).

In general, a buck-and-boost transformer is a four-winding isolated transformer designed so that the independent windings may be interconnected to function as an autotransformer. Connected in this manner, all power for the load must pass through the transformer windings as shown in Fig. 3-4. Then, by proper interconnection of the windings, the output voltage may be increased or decreased from the input voltage depending on the ratio between the primary and secondary windings.

Connected as an autotransformer, the largest part of the power goes directly to the load, with only that part bucking or boosting being involved in transformation (see Fig. 3-5). If, say, only 10% of the voltage must undergo transformation, the autotransformer rating is increased approximately 10 times the normal rating of the corresponding isolated transformer.

An approximate transformer size can be determined by multiplying the volts (buck or boost) times the load amperes. This can only be an approximation, due to the wide requirement for voltage change and the amperage used. It is possible for this method to exceed the transformer rating.

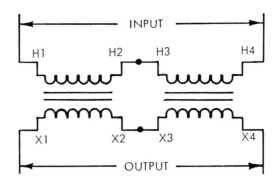

Figure 3-4 Four-winding isolated transformer. (Courtesy Borg-Warner Air Conditioning.)

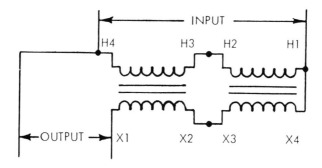

Figure 3-5 Autotransformer. (Courtesy Borg-Warner Air Conditioning.)

CONTROL CIRCUIT TRANSFORMERS

Low-voltage control circuit transformers (class 2) are used extensively in control circuits to obtain a lower voltage than is available from the power supply. For example, many control circuits operate at 24 V, and normally 120 V is the lowest voltage used in any electrical system for building construction. Therefore, a transformer is used to reduce the 120-V circuit to the required 24 V. In selecting such a transformer, class 2 low-voltage control systems are limited to transformers with a maximum output capacity of 75 VA (watts). If a control transformer is overloaded for any significant length of time, the transformer will fail. Therefore, systems that require the addition of controls should be checked to assure that the rating of the transformer will not be exceeded. A typical control circuit is shown in Fig. 3-6, showing the load of the holding coils, which totals 22 VA. Since a 45-VA transformer is used, (45 minus 22 =) 23 VA is available for additional field-installed controls.

MOTOR STARTERS

As mentioned previously, a motor starter provides a means to make and break all power supply lines running to a motor and also provides motor overload protection. A typical motor starter is shown in Fig. 3-7.

The main purpose of the overload protection is to protect the motor and its controls against overheating due to motor overloads. Such protection should be selected to trip at not more than 125% full-load motor current for mermetic compressor motors and motors with a temperature rise of not over 40°C. When the value specified for motor overcurrent protection does not correspond to a standard-size protector, the next higher rating (but not greater than 140% of full-load current rating) may be used.

For all other motors, the trip should be selected at not more than 115% full-

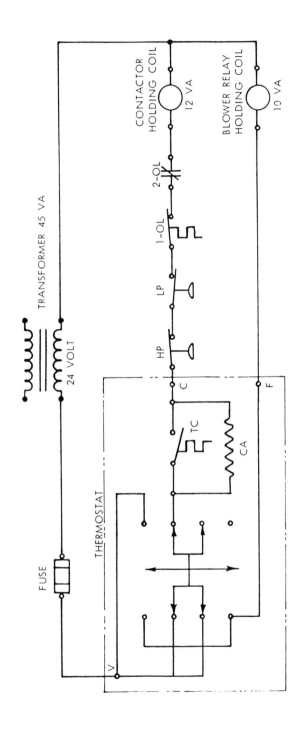

TRANSFORMER CAPACITY - 45 VA

LOAD IN CIRCUIT - 12 (CONTACTOR HOLDING COIL) PLUS 10 (RELAY HOLDING COIL) = 22

45 - 22 = 23 VA AVAILABLE FOR ADDITIONAL FIELD INSTALLED CONTROLS

Figure 3-6 Typical control circuit: transformer loading. (Courtesy Borg-Warner Air Conditioning.)

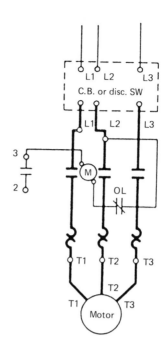

Figure 3-7 Typical motor starter. (Courtesy Borg-Warner Air Conditioning.)

load motor current, except where the value specified does not correspond to a standard-size protector. Then the next higher rating (but not greater than 130% of full-load current rating) may be used.

The motor starter is also rated in amperes or horsepower and must be selected with a capacity rating greater than the load in which it serves. If a motor load is only slightly less than the rated capacity of the starter, a longer contact life can be obtained by selecting the next-size-larger starter.

Motor starters are normally available with two-, three-, or four-pole contacts. The holding coil may be for use with line-voltage or 24-V control circuits. The schematic drawing in Fig. 3-8 will serve to demonstrate the operation of a motor starter. Starting at the top of the drawing, note that the power circuit is protected with a fused disconnect switch and that three-phase is used to power the motor. This power supply runs through the motor starter prior to connecting to the motor itself. Note that this particular starter has four poles, but only three are being used.

The drawing in Fig. 3-8 represents a control circuit for an air-conditioning system. The holding coil, which controls making and breaking the contacts, is connected in series with the thermostat and safety controls. When the thermostat closes and all safety control contacts are closed, the holding coil is energized, closing the starter contacts and supplying power to the compressor motor. When the thermostat is satisfied, its contact opens to deenergize the holding coil and interrupt the power supply to the compressor. If any one of the power legs should exceed the current rating of the overload element, the corresponding overload contact will open

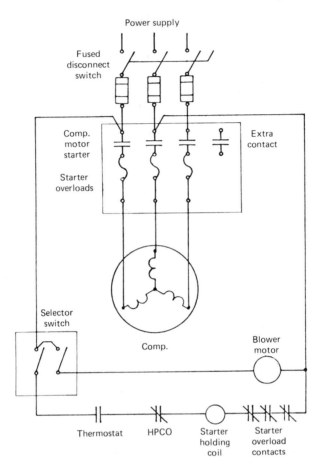

Figure 3-8 Typical system wiring diagram with four-pole motor starter. (Courtesy Borg-Warner Air Conditioning.)

to deenergize the holding coil. The extra starter contact in this control could be used for control of a remote motor, such as a cooling tower pump or the like.

The type, size, and manufacturer of the starter will vary the VA requirement of the holding coil. Typical holding coil ratings are as follows:

1. *30-A:* 9 to 11 VA
2. *60-A:* 17 to 20 VA

To select a motor starter, in general, first obtain the motor characteristics from manufacturers' data or look at the nameplate on the motor. For example, assume a 5-hp motor with a full-load ampere rating of 17.4 at 240 V. After checking in a typical motor control catalog, you would choose a NEMA size 1 starter (for a 5-hp motor). Overload elements are then selected to trip at 21.8 A (1.25 × 17.4). The holding coil could be rated at full line voltage or 240 V. If the motor is designed for three-phase operation, at least a three-pole starter will be utilized, or a two-pole starter will suffice for a single-phase motor.

CONTACTORS

The purpose of contactors is to provide a means to make and break all power supply lines running to a load. In the case of electric motors, the load will be inductive. The capacity rating of contactors must be selected with a rating greater than the load it serves. If a load is only slightly less than the rated capacity of the contactor, a longer contact life can be obtained by selecting the next-size-larger contactor and is recommended when specifying definite-purpose devices.

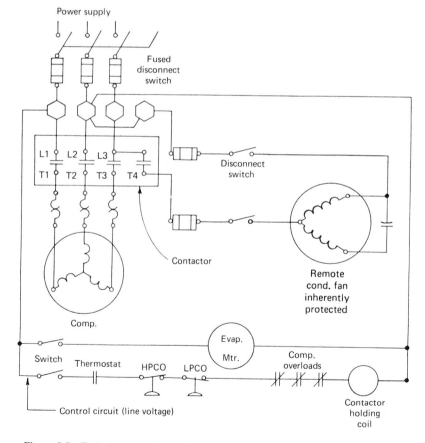

Figure 3-9 Typical system wiring diagram with four-pole contactor. (Courtesy Borg-Warner Air Conditioning.)

The contactor holding coil may be for either line voltage or low-voltage control circuits, depending on the application. Starters are available with two, three, or four contacts for various applications. For example, refer to Fig. 3-9, where a four-pole contactor is utilized. Note that the contacts are connected in series with the power supply to the load. The holding coil, which controls the making and breaking of the contacts, is connected in series with the thermostat and safety controls in the control circuit. When the thermostat closes and all safety control contacts are closed, the holding coil is energized, closing the contacts of the contactor to supply power to the load. Once the thermostat is satisfied, its contact opens to deenergize the holding coil, which interrupts the power supply to the load. The extra, fourth contact is used to control the power supply to a small remote motor operating a condenser fan. The type, size, and manufacturer of the contactor vary with the VA requirement of the holding coil, typically rated in the range 9 to 20 VA.

PROTECTOR RELAY (LOCKOUT) CONTROL CIRCUITS

The purpose of the lockout relay in a control circuit is to prevent a certain motor from operating out of sequence due to the opening and closing of the control circuit by any of the automatic reset safety controls. If the lockout relay has been activated, the system may be reset for normal operation only by interrupting the power supply to the control circuit either by the thermostat (in the case of a HVAC system) or the main power switch.

In the circuit in Fig. 3-10, the coil, due to its high resistance, will not be energized during normal operation. However, when any one of the safety controls opens the circuit to the compressor contactor coil, current flows through the lockout relay coil, causing the coil to become energized and to open its contact. This contact will then remain open, keeping the compressor contactor circuit open until the power is interrupted either by the thermostat or by the main power switch after the safety control has reset. Performance depends on the resistance of the lockout relay coil being much greater than the resistance of the compressor contactor coil. If the lockout relay becomes defective, it should be replaced with an exact duplicate to maintain the proper resistance balance.

It is permissible to add a control relay coil in parallel with the contactor coil when a system demands another control. The resistance of the contactor coil and the relay coil in parallel will decrease the total resistance and will not affect the operation of the lockout relay. However, never put additional lockout relays, lights, or other load devices in parallel with the lockout relay coil.

TIME-DELAY RELAYS

The purpose of the time-delay relay is to delay, for a predetermined length of time after the control system has been energized, the normal operation of a control or

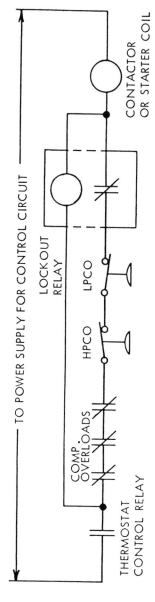

Figure 3-10 Typical control circuit with lockout relay. (Courtesy Borg-Warner Air Conditioning.)

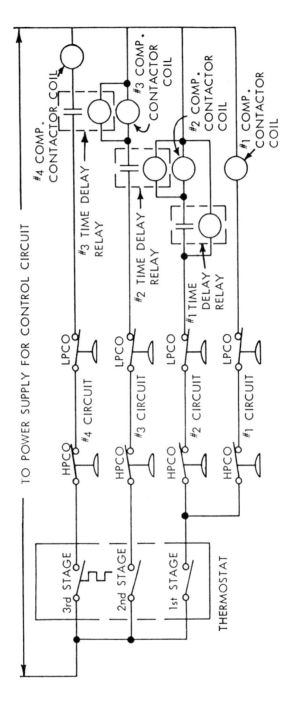

Figure 3-11 Typical four-compressor control circuit with time-delay relays for sequencing compressor starting. (Courtesy Borg-Warner Air Conditioning.)

group of controls. The length of delay depends on the time built into the relay coil and may vary from a fraction of a second to several minutes.

Electrical systems utilizing several motors may use such a delay device to start the motors one at a time to limit inrush current. For example, the schematic drawing in Fig. 3-11 shows four compressor control circuits with time-delay relays for sequencing the starting of each. In this design, if all three stages of the thermostat are closed and electrical power is supplied to the units, the #1 compressor contactor coil and the #1 time-delay coil will become energized to start the #1 compressor. After the specified time delay, the contacts of the #1 time-delay relay will close to energize the #2 compressor contactor coil and the #2 time-delay coil, which starts the #2 compressor. After the specified time delay, the contacts of the #2 time-delay relay will close to energize the #3 compressor contactor coil and the #3 time-delay coil, which starts the #3 compressor, and so on, until all four compressors are in operation.

Another application of time-delay relays is on HVAC systems with air-cooled condensers operating with zero-degree low-ambient accessories. A time-delay relay may be used, for example, to bypass the automatic low-pressure cutout for a sufficient length of time to permit the suction pressure to build up and close the contacts of the cutout.

DISCONNECT SWITCHES

Every motor circuit must have a disconnect switch to disconnect the power supply from the motor for complete shutdown or servicing of the unit. The disconnect also provides a means of overcurrent protection for the electrical conductors and controls in the circuit.

Each motor must have a disconnect switch, and this switch must be readily accessible to the load it serves. Disconnect switches are rated in amperes and horsepower, and each should be selected with a capacity rating greater than the load it serves. The recommended switch rating is 125% of the full-load motor current. Therefore, if a motor has a full-load ampere rating of, say, 20.5 A, the switch should be rated at $20.5 \times 1.25 = 25.62$ A. The smallest disconnect switch now made is rated for 30 A, so this would suffice nicely for this load.

When circuit protection is not required at the motor location, nonfusible disconnects are available at a lower price than that of the fusible types. Two-pole disconnects are used on single-phase circuits and three-pole disconnects are used for three-phase circuits.

When circuit protection is desired, fuses are utilized to protect the electrical circuit from damage due to high or dangerous temperatures—resulting from excessive current. Two types are common: the plug fuse and the cartridge fuse.

The *plug fuse* is rated only to 30 A and may be used only on very small motors. These are also rated to only 125 V and screw into a conventional Edison-base full holder.

Several types of *cartridge fuses* are currently in use. The two major physical characteristics are the contact fuse with end caps and the type with blades. The type with end caps is rated from 0 to 60 A and from 0 to 600 V. The blade types are rated from about 70 to 600 A and from 0 to 600 V.

In addition to their outside physical characteristics, cartridge fuses differ internally. The one-time cartridge fuse has a permanent connected fuse link and when this link opens, the link must be discarded. Renewable link fuses are similar, but they have replaceable fuse links. Therefore, when the link opens, it may be replaced with a new one.

For motor applications, the dual-element fuse (either plug or cartridge type) is the recommended type in most cases. This type of fuse has a fusible link and also a thermal cutout connected in series. The fusible link protects against short circuits, while the thermal cutout protects against overloads. When a motor is started, higher amperage occurs in the circuit than during normal running operation. Therefore, a long-time lag built into the fuse prevents opening the circuit on a normal motor startup.

Fuse ratings are based on current that trips the thermal cutout. Their use permits selecting a smaller fuse size to afford close overload protection, yet provides trip-free motor starting current. The NEC and manufacturers' catalogs provide the recommended maximum fuse sizes for each application.

SOLENOID VALVES

A solenoid valve is used in liquid lines and is electrically opened or closed to control the flow of the fluid. It is commonly used in manufacturing processes, in the liquid line to an evaporator to control the flow of refrigerant in an HVAC system, and in other applications.

In most cases, the solenoid coil opens the valve when energized and allows the valve to close when deenergized. They are almost always operated at line voltage since the VA rating is normally too large for a 24-V control system.

When the control circuit is energized, the relay contacts close to supply power to the solenoid coil, which causes the valve to open. When the control circuit is deenergized, the relay contacts open to interrupt power supplying the coil, causing the valve to close.

In selecting solenoid valves, the following must be determined:

1. Fluid to be controlled
2. Quantity of fluid the valve must allow to flow
3. Allowable pressure drop
4. Maximum operating pressure differential
5. Maximum working pressure
6. Electrical characteristics for coil operation

TABLE 3-1   VA Capacity of Control Circuits: Copper Conductors Based on
Recommended 3% Voltage Drop

AWG Wire Size	Length of Circuit, One Way (ft)											
	25	50	75	100	125	150	175	200	225	250	275	300
20	29	14	10	7.2	5.8	4.8	4.1	3.6	3.2	2.9	2.6	2.4
18	58	29	19	14	11	9.6	8.2	7.2	6.4	5.8	5.2	4.8
16	86	43	29	22	17	14	12	11	9.6	8.7	7.8	7.2
14	133	67	44	33	27	22	19	17	15	13	12	11

Source: Borg-Warner Air Conditioning.

When low-voltage lines are installed, it is suggested that one extra line be run for emergency purposes. This can be substituted for any one of the existing lines that may be defective. Also, it is possible to parallel this extra line with the existing line carrying the full load of the control circuit if the length of run affects control operation caused by voltage drop. In many cases this will reduce the voltage drop and permit satisfactory operation.

CONTROL RELAYS

In general, the purpose of a control relay is to energize or deenergize an electrical circuit to obtain a specific operation of a component. It may be used to control a motor, heater, solenoid valve, or another relay.

In selecting a control relay, determine the voltage to be applied to the coil. Then determine the voltage and current characteristics of the load to be controlled by the contacts and whether the load is resistive or inductive. Determine the coil VA rating (this is especially important on low-voltage systems) and then consult a manufacturer's catalog to select a relay to meet the requirements.

chapter four

Electric Motors

Any study of electric motor controls should include a brief review of electric motor principles. In basic terms, electric motors convert electric energy into the productive power of rotary mechanical force. This capability finds application in unlimited ways: from explosion-proof, water-cooled motors for underground mining to induced-draft fan motors for power generation; from adjustable-frequency drives for waste and water treatment pumping to eddy-current clutches for automobile production; from direct-current drive systems for paper production to photographic film manufacturing; from rolled-shell shaftless motors for machine tools to large outdoor motors for crude oil pipelines; from mechanical variable-speed drives for woodworking machines to complex adjustable-speed drive systems for textiles. All these and more represent the scope of electric motor participation in powering and controlling the machines and processes of industries throughout the world.

Before covering electric motor operating principles and their applications, certain motor terms must be understood. Some of the more basic ones are as follows.

Style number. Identifies that particular motor in contrast to all others. Manufacturers provide style numbers on the motor nameplate and in the written specifications.

Serial data code. The first letter is a manufacturing code used at the factory. The second letter identifies the month, and the last two numbers identify the year of manufacture (D78 is April 78).

Frame.   Specifies the shaft height and motor mounting dimensions and provides recommendations for standard shaft diameters and usable shaft extension lengths.

Service factor.   A service factor (SF) is a multiplier that when applied to the rated horsepower, indicates a permissible horsepower loading that may be carried continuously when the voltage and frequency are maintained at the value specified on the nameplate, although the motor will operate at an increased temperature rise.

NEMA service factors.   Open motors only:

Hp	SF	Hp	SF	Hp	SF
$\frac{1}{12}$	1.40	$\frac{1}{3}$	1.35	1	1.15
$\frac{1}{8}$	1.40	$\frac{1}{2}$	1.25	$1\frac{1}{2}$	1.15
$\frac{1}{6}$	1.35	$\frac{3}{4}$	1.25	2	1.15
$\frac{1}{4}$	1.35			3	1.15

Phase.   Indicates whether the motor has been designed for single- or three-phase operation. It is determined by the electrical power source.

Degree C ambient.   The air temperature immediately surrounding the motor. Forty degrees Celsius is the NEMA maximum ambient temperature.

Insulation class.   The insulation system is chosen to ensure that the motor will perform at the rated horsepower and service factor load.

Horsepower.   Defines the rated output capacity of the motor. It is based on breakdown torque, which is the maximum torque that a motor will develop without an abrupt drop in speed.

Rpm.   Revolutions per minute (speed). The rpm reading on motors is the approximate full-load speed. The speed of the motor is determined by the number of poles in the winding. A two-pole motor runs at an approximate speed of 3450 rpm. A four-pole motor runs at an approximate speed of 1725 rpm. A six-pole motor runs at an approximate speed of 1140 rpm.

Amperes.   Gives the amperes of current the motor draws at full load. When two values are shown on the nameplate, the motor usually has a dual voltage rating. Volts and amperes are inversely proportional; the higher the voltage, the lower the amperes, and vice versa. The higher ampere value corresponds to the lower voltage rating on the nameplate. Two-speed motors will also show two ampere readings.

Hertz (cycles).   Just about everything in this country is serviced by 60-Hz alternating current. Therefore, most applications will be for 60-Hz operations.

Volts.   Volts is the electrical potential "pressure" for which the motor is designed. Sometimes two voltages are listed on the nameplate, such as 115/230. In this case the motor is intended for use on either a 115-V or a 230-V circuit. Special instructions are furnished for connecting the motor for each voltage.

KVA code.   This code letter is defined by NEMA standards to designate the locked rotor kilovolt-amperes (kVA) per horsepower of a motor. It relates to starting current and selection of fuse or circuit breaker size.

Housing.   Designates the type of motor enclosure. The most common types are open and enclosed:

- *Open drip-proof:* Has ventilating openings so constructed that successful operation is not interfered with when drops of liquid or solid particles strike or enter the enclosure at any angle from 0 to 15° downward from vertical.
- *Open guarded:* Has all openings giving direct access to live metal hazardous rotating parts so sized or shielded as to prevent accidental contact as defined by probes illustrated in the NEMA standard.
- *Totally enclosed:* Motors are so constructed as to prevent the free exchange of air between the inside and outside of the motor casing.
- *Totally enclosed fan-cooled:* Motors are equipped for external cooling by means of a fan that is integral with the motor.
- *Air-over:* Motors must be mounted in the airstream to obtain their nameplate rating without overheating. An air-over motor may be either open or enclosed.

Explosion-proof motors.   These are totally enclosed designs built to withstand an explosion of gas or vapor within it, and to prevent ignition of the gas or vapor surrounding the motor by sparks or explosions that may occur within the motor casing.

Hours.   Designates the duty cycle of a motor. Most fractional-horsepower motors are marked continuous for around-the-clock operation at the nameplate rating in the rated ambient. Motors marked "one-half" are for one-half-hour ratings, and those marked "one" are for one-hour ratings.

The following terms are not found on the nameplate but are important considerations for proper motor selection.

Sleeve bearings.   Sleeve bearings are generally recommended for axial thrust loads of 210 lb or less and are designed to operate in any mounting position as long as the belt pull is not against the bearing window. On light-duty applications, sleeve bearings can be expected to perform a minimum of 25,000 hours without relubrication.

Ball bearings.   These are recommended where axial thrust exceeds 20 lb. They, too, can be mounted in any position. Standard and general-purpose ball-bearing motors are factory lubricated and under normal conditions will require no additional lubrication for many years.

Rigid mounting.   A rectangular steel mounting plate that is welded to the motor frame or cast integral with the frame; it is the most common type of mounting.

Resilient mounting.   A mounting base that is isolated from motor vibration by means of rubber rings secured to the end bells.

Flange mounting.   A special end bell with a machined flange that has two or more holes through which bolts are secured. Flange mountings are commonly used on such applications as jetty pumps and oil burners.

Rotation.   For single-phase motors, the standard rotation, unless otherwise noted, is counterclockwise facing the lead or opposite shaft end. All motors can be reconnected at the terminal board for opposite rotation, unless otherwise indicated.

SINGLE-PHASE MOTORS

Split-Phase Motors

Split-phase motors are fractional-horsepower units that use an auxiliary winding on the stator to aid in starting the motor until it reaches its proper rotation speed (see Fig. 4-1). This type of motor finds use in small pumps, oil burners, automatic washers, and other household appliances.

In general, the split-phase motor consists of a housing, a laminated iron-core stator with embedded windings forming the inside of the cylindrical housing, a rotor made up of copper bars set in slots in an iron core and connected to each other by copper rings around both ends of the core, plates that are bolted to the housing and contain the bearings that support the rotor shaft, and a centrifugal switch inside the housing. This type of rotor is often called a *squirrel-cage rotor* since the configuration of the copper bars resembles a cage. These motors have no windings, as such, and a centrifugal switch is provided to open the circuit to the starting winding when the motor reaches running speed.

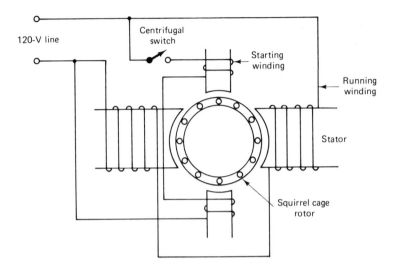

Figure 4-1 Wiring diagram of a split-phase motor.

To understand the operation of a split-phase motor, look at the wiring diagram in Fig. 4-1. Current is applied to the stator windings, both the main winding and the starting winding, which is in parallel with it through the centrifugal switch. The two windings set up a rotating magnetic field, and this field sets up a voltage in the copper bars of the squirrel-cage rotor. Because these bars are shortened at the ends of the rotor, current flows through the rotor bars. The current-carrying rotor bars then react with the magnetic field to produce motor action. When the rotor is turning at the proper speed, the centrifugal switch cuts out the starting winding since it is no longer needed.

Capacitor Motors

Capacitor motors are single-phase ac motors ranging in size from fractional horsepower to perhaps as high as 15 hp. This motor is widely used in all types of single-phase applications, such as powering machine shop tools (lathes, drill presses, etc.), air compressors, refrigerators, and the like, and it is similar in construction to the split-phase motor, except that a capacitor is wired in series with the starting winding, as shown in Fig. 4-2.

The capacitor provides higher starting torque, with lower starting current, than does the split-phase motor, and although the capacitor is sometimes mounted inside the motor housing, it is usually mounted on top of the motor and is encased in a metal compartment.

In general, two types of capacitor motors are in use: the capacitor-start motor and the capacitor-start-and-run motor. As the name implies, the *capacitor-start*

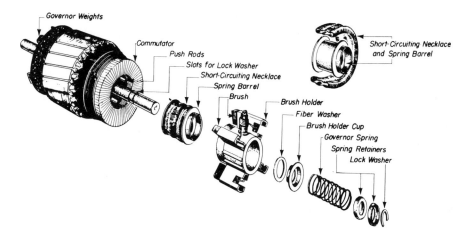

Figure 4-2 Rotor assembly of a repulsion-start motor.

motor utilizes the capacitor only for starting; it is disconnected from the circuit once the motor reaches running speed, or at about 75% of the motor's full speed. Then the centrifugal switch opens to cut the capacitor out of the circuit.

The *capacitor-start-and-run motor* keeps the capacitor and starting winding in parallel with the running winding, providing quiet and smooth operation at all times.

Capacitor split-phase motors require the least maintenance of all single-phase motors, but they have a very low starting torque, making them unsuitable for many applications. Their high maximum torque, however, makes them especially useful for such tools as floor sanders or in grinders, where momentary overloads due to excessive cutting pressure are experienced. They are also frequently used for slow-speed direct-connected fans.

Repulsion-Type Motors

Repulsion-type motors are divided into several groups, including (1) repulsion-start, induction-run motors, (2) repulsion motors, and (3) repulsion-induction motors. The *repulsion-start, induction-run motor* is of the single-phase type, ranging in size from about $\frac{1}{10}$hp to as high as 20 hp. It has high starting torque and a constant-speed characteristic, which makes it suitable for such applications as commercial refrigerators, compressors, pumps, and similar applications requiring high starting torque.

The *repulsion motor* is distinguished from the repulsion-start, induction-run motor by the fact that it is made exclusively as a brush-riding type and does not have any centrifugal mechanism. Therefore, this motor both starts and runs on the repulsion principle. It has high starting torque and a variable-speed characteristic. It is reversed by shifting the brush holder to either side of the neutral position. And its speed can be decreased by moving the brush holder farther away from the neutral position.

The *repulsion-induction motor* combines the high starting torque of the repulsion-type and the good speed regulation of the induction motor. The stator of this motor is provided with a regular single-phase winding, and the rotor winding is similar to that used on a dc motor. When starting, the changing single-phase stator flux cuts across the rotor windings and induces currents in them; thus, when flowing through the commutator, a continuous repulsive action on the stator poles is present.

This motor starts as a straight repulsion-type and accelerates to about 75% of normal full speed with a centrifugally operated device that connects all the commutator bars together and converts the winding to an equivalent squirrel-cage type. The same mechanism usually raises the brushes to reduce noise and wear. Note that when the machine is operating as a repulsion-type, the rotor and stator poles reverse at the same instant and that the current in the commutator and brushes is ac.

This type of motor will develop four to five times normal full-load torque and will draw about three times normal full-load current when starting with full-line voltage applied. The speed variation from no load to full load will not exceed 5% of normal full-load speed.

The repulsion-induction motor is used to power air compressors, refrigerators, freezers, pumps, meat grinders, small lathes, small conveyors, stokers, and the like. In general, the motor is suitable for any load that requires a high starting torque and constant-speed operation. Most motors of this type are less than 5 hp.

Universal Motors

This motor is a special adaptation of the series-connected dc motor, and it gets its name "universal" from the fact that it can be connected on either ac or dc and will operate the same. It is a single-phase motor for use on 120 or 240 V.

In general, the universal motor contains field windings on the stator within the frame, an armature with the ends of its windings brought out to a commutator at one end, and carbon brushes that are held in place by the motor's end plate, allowing them to make a proper contact with the commutator.

When current is applied to a universal motor, either ac or dc, the current flows through the field coils and the armature windings in series. The magnetic field set up by the field coils in the stator react with the current-carrying wires on the armature to produce rotation. Universal motors are used on such household appliances as sewing machines, vacuum cleaners, and electric fans.

Shaded-Pole Motors

A shaded-pole motor is a single-phase induction motor with an uninsulated and permanently short-circuited auxiliary winding displaced in magnetic position

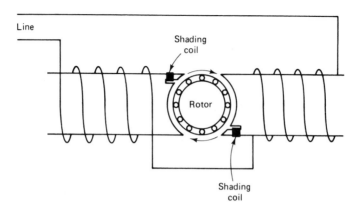

Figure 4-3 Diagram of a shaded-pole motor.

from the main winding. The auxiliary winding is known as the shading coil and usually surrounds from one-third to one-half of the pole (see Fig. 4-3). The main winding surrounds the entire pole and may consist of one or more coils per pole.

Applications for this motor include small fans, timing devices, relays, radio dials, or any constant-speed load not requiring high starting torque.

POLYPHASE MOTORS

Three-phase motors offer extremely efficient and economical application and are the motors usually preferred for commercial and industrial applications when three-phase service is available. In fact, almost all motors sold are standard ac three-phase motors. These motors are available in ratings from fractional horsepower up to thousands of horsepower in almost every standard voltage and frequency. In fact, there are few applications for which the three-phase motor cannot be put to use.

Three-phase motors are known for their relatively constant speed characteristic and are available in designs giving a variety of torque characteristics; that is, some have a high starting torque and others have a low starting torque. Some are designed to draw a normal starting current and others to draw a high starting current.

A typical three-phase motor is shown in Fig. 4-4. Note that the three main parts are the stator, rotor, and end plates. This motor is very similar in construction to conventional split-phase motors except that the three-phase motor has no centrifugal switch.

The stator shown in Fig. 4-5 consists of a steel frame and a laminated iron core and winding formed of individual coils placed in slots. The rotor may be of squirrel-cage or wound-rotor type. Both types contain a laminated core pressed onto a shaft. The squirrel-cage rotor is shown in Fig. 4-6 and is similar to a split-phase

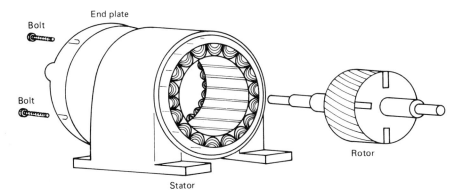

Bolt

End plate

Bolt

Rotor

Stator

Figure 4-4 Typical three-phase motor in an exploded view. (Courtesy Marathon Electric, Inc.)

motor. The wound rotor is shown in Fig. 4-7 and has a winding on the core that is connected to three slip rings mounted on the shaft. The end plates or brackets are bolted to each side of the stator frame and contain the bearings in which the shaft revolves. Either ball bearings or sleeve bearings are used.

Figure 4-5 Stator of a three-phase motor. (Courtesy Dreisilker Electric Motors, Inc.)

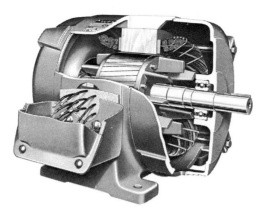

Figure 4-6   Squirrel-cage rotor. (Courtesy Westinghouse.)

Induction motors, both single-phase and polyphase, get their name from the fact that they utilize the principle of electromagnetic induction. An induction motor has a stationary part, or stator, with windings connected to the ac supply, and a rotation part, or rotor, which contains coils or bars. There is no electrical connection

Figure 4-7   Wound rotor for three-phase motor. (Courtesy Dreisilker Electric Motors, Inc.)

between the stator and rotor. The magnetic field produced in the stator windings induces a voltage in the rotor coils or bars.

Since the stator windings act in the same way as the primary winding of a transformer, the stator of an induction motor is sometimes called the primary. Similarly, the rotor is called the secondary because it carries the induced voltage in the same way as the secondary of a transformer.

The magnetic field necessary for induction to take place is produced by the stator windings. Therefore, the induction-motor stator is often called the *field* and its windings are called *field windings*. The terms *primary* and *secondary* relate to the electrical characteristics and the terms *stator* and *rotor* to the mechanical features of induction motors. The rotor transfers the rotating motion to its shaft, and the revolving shaft drives a mechanical load or a machine, such as a pump, spindle, or clock.

Commutator segments, which are essential parts of dc motors, are not needed on induction motors. This greatly simplifies the design and the maintenance of induction motors as compared to dc motors.

The turning of the rotor in an induction motor is due to induction. The rotor, or secondary, is not connected to any source of voltage. If the magnetic field of the stator, or primary, revolves, it will induce a voltage in the rotor, or secondary. The magnetic field produced by the induced voltage acts in such a way that it makes the secondary follow the movement of the primary field.

The stator, or primary, of the induction motor does not move physically. The movement of the primary magnetic field must thus be achieved electrically. A rotating magnetic field is made possible by a combination of two or more ac voltages that are out of phase with each other and applied to the stator coils. Direct current will not produce a rotating magnetic field. In three-phase induction motors, the rotating magnetic field is obtained by applying a three-phase system to the stator windings.

The direction of rotation of the rotor in an ac motor is the same as that of its rotating magnetic field. In a three-phase motor the direction can be reversed by interchanging the connections of any two supply leads. This interchange will reverse the sequence of phases in the stator, the direction of the field rotation, and therefore the direction of rotor rotation.

SYNCHRONOUS MOTORS

A synchronous polyphase motor has a stator constructed in the same way as the stator of a conventional induction motor. The iron core has slots into which coils are wound, which are also arranged and connected in the same way as the stator coils of the induction motor. These are in turn grouped to form a three-phase connection, and the three free leads are connected to a three-phase source. Frames are equipped with air ducts, which aid the cooling of the windings, and coil guards protect the winding from damage.

The rotor of a synchronous motor carries poles that project toward the armature; they are called *salient poles*. The coils are wound on laminated pole bodies and connected to slip rings on the shaft. A squirrel-cage winding for starting the motor is embedded in the pole faces.

The pole coils are energized by direct current, which is usually supplied by a small dc generator called the *exciter*. This exciter may be mounted directly on the shaft to generate dc voltage, which is applied through brushes to slip rings. On low-speed synchronous motors, the exciter is normally belted or of a separate high-speed motor-driven type.

The dimensions and construction of synchronous motors vary greatly depending on the rating of the motors. However, synchronous motors for industrial power applications are rarely built for less than 25 hp or so. In fact, most are 100 hp or more. All are polyphase motors when built in this size. Vertical and horizontal shafts with various bearing arrangements and various enclosures cause wide variations in the appearance of the synchronous motor.

Synchronous motors are used in electrical systems where there is need for improvement in power factor or where low power factor is not desirable. This type of motor is especially adapted to heavy loads that operate for long periods of time without stopping, such as for air compressors, pumps, ship propulsion, and the like.

The construction of the synchronous motor is well adapted for high voltages, as it permits good insulation. Synchronous motors are frequently used at 2300 V or more. Its efficient slow-running speed is another advantage.

DIRECT-CURRENT MOTORS

A direct-current motor is a machine for converting dc electrical energy into rotating mechanical energy. The principle underlying the operation of a dc motor is called *motor action* and is based on the fact that when a wire carrying current is placed in a magnetic field, a force is exerted on the wire, moving it through the magnetic field. There are three elements to motor action as it takes place in a dc motor:

1. Many coils of wire are wound on a cylindrical rotor or armature on the shaft of the motor.
2. A magnetic field necessary for motor action is created by placing fixed electromagnetic poles around the inside of the cylindrical motor housing. When current is passed through the fixed coils, a magnetic field is set up without the housing. Then, when the armature is placed inside the motor housing, the wires of the armature coils will be situated in the field of magnetic lines of force set up by the electromagnetic poles arranged around the stator. The stationary cylindrical part of the motor is called the *stator*.
3. The shaft of the armature is free to rotate because it is supported at both ends by bearing brackets. Freedom of rotation is assured by providing clearance between the rotor and the faces of the magnetic poles.

Shunt-Wound DC Motors

In this type of motor, since the strength of the field is not affected appreciably by a change in the load, a relatively constant speed is obtainable. This type of motor may be used for the operation of machines that require an approximate constant speed and impose low starting torque and light overload on the motor.

Series-Wound DC Motors

In motors of this type, any increase in load results in more current passing through the armature and the field windings. As the field is strengthened by this increased current, the motor speed decreases. Conversely, as the load is decreased, the field is weakened and the speed increases, and at very light loads speed may become excessive. For this reason, series-wound motors are usually directly connected or geared to the load to prevent runaway. Because the increase in armature current with an increasing load produces increased torque, the series-wound motor is especially suited to heavy starting duty and where severe overloads may be expected. Its speed may be adjusted by means of a variable resistance placed in series with the motor, but due to variation with load, the speed cannot be held at any constant value. This variation of speed with load becomes greater as the speed is reduced. Use of this motor is normally limited to traction and lifting service.

Compound-Wound Motors

In this type of motor, the speed variation due to the load changes is much less than in the series-wound motor but is greater than in the shunt-wound motor. It also has a greater starting torque than the shunt-wound motor and is able to withstand heavier overloads. However, it has a narrower adjustable-speed range. Standard motors of this type have a cumulative-compound winding, the differential-compound winding being limited to special applications. They are used where the starting load is very heavy or where the load changes suddenly and violently, as with reciprocating pumps, printing presses, and punch presses.

Brushless DC Motors

The brushless dc motor was developed to eliminate commutator problems in missiles and spacecraft in operation above the earth's atmosphere. Two general types of brushless motors are in use: the inverter-induction motor and a dc motor with an electronic commutator.

The *inverter-induction motor* has an inverter that uses the motor windings as the usual filter. The operation is square wave, and the combined efficiencies of the inverter and induction motor are at least as high as for a dc motor alone. In all cases, the motors must be designed to saturate so that starting current is limited; otherwise, the transistors or silicon-controlled rectifiers in the inverter will be overloaded.

MOTOR ENCLOSURES

Electric motors differ in construction and appearance depending on the type of service for which they are to be used. Open and closed frames are common. In the former enclosure, the motor's parts are covered for protection, but the air can freely enter the enclosure. Further designations for this type of enclosure include drip-proof, weather-protected, and splash-proof.

Totally enclosed motors, such as the one shown in Fig. 4-4, have an airtight enclosure. They may be fan cooled or self-ventilated. An enclosed motor equipped with a fan has the fan as an integral part of the machine but it is external to the enclosed parts. In the self-ventilated enclosure, no external means of cooling is provided.

The type of enclosure to use will depend on the ambient and surrounding conditions. In a drip-proof machine, for example, all ventilating openings are constructed so that drops of liquid or solid particles falling on the machine at an angle of not greater than 15° from the vertical cannot enter the machine, even directly or by striking and running along a horizontal or inclined surface of the machine. The application of this machine would lend itself to areas where liquids are processed.

An open motor having all air openings that give direct access to live or rotating parts, other than the shaft, limited in size by the design of the parts or by screen to prevent accidental contact with such parts is classified as a drip-proof, fully guarded machine. In such enclosures, openings must not permit the passage of a cylindrical rod $\frac{1}{2}$ in. in diameter, except where the distance from the guard to the live rotating parts is more than 4 in., in which case the openings must not permit the passage of a cylindrical rod $\frac{3}{4}$ in. in diameter.

There are other types of drip-proof machines for special applications, such as externally ventilated and pipe ventilated, which as the names imply, are either ventilated by a separate motor-driven blower or cooled by ventilating air from inlet ducts or pipes.

An enclosed motor whose enclosure is designed and constructed to withstand an explosion of a specified gas or vapor that may occur within the motor and to prevent the ignition of this gas or vapor surrounding the machine is designated an "explosion-proof" (XP) motor.

Hazardous atmospheres (requiring XP enclosures) of both a gaseous and dusty nature are classified by the NEC as follows:

- *Class I, group A:* atmospheres containing acetylene
- *Class I, group B:* atmospheres containing hydrogen gases or vapors of equivalent hazards such as manufactured gas
- *Class I, group C:* atmospheres containing ethyl ether vapor
- *Class I, group D:* atmospheres containing gasoline, petroleum, naphtha, alcohols, acetone, lacquer-solvent vapors, and natural gas

TABLE 4-1 NEMA Designs for Polyphase Motors

NEMA Design	Starting Torque	Starting Current	Breakdown Torque	Full-Load Slip
A	Normal	Normal	High	Low
B	Normal	Low	Medium	Low
C	High	Low	Normal	Low
D	Very high	Low	—	High

- *Class II, group E:* atmospheres containing metal dust
- *Class II, group F:* atmospheres containing carbon-black, coal, or coke dust
- *Class II, group G:* atmospheres containing grain dust

The proper motor enclosure must be selected to fit the particular atmosphere. Explosion-proof equipment is not generally available for class I, groups A and B, and thus it is necessary to isolate motors from the hazardous area.

MOTOR TYPE

The type of motor will determine the electrical characteristics of the design. NEMA-designated designs for polyphase motors are given in Table 4-1.

An A motor is a three-phase squirrel-cage motor designed to withstand full-voltage starting with locked rotor current higher than the values for a B motor and having a slip at rated load of less than 5%.

A B motor is a three-phase squirrel-cage motor designed to withstand full-voltage starting and developing locked rotor and breakdown torques adequate for general application, and having a slip at rated load of less than 5%.

A C motor is a three-phase squirrel-cage motor designed to withstand full-voltage starting, developing locked rotor torque for special high-torque applications, and having a slip at rated load of less than 5%.

A D motor is a three-phase squirrel-cage motor designed to withstand full-voltage starting, developing 275% locked rotor torque, and having a slip at rated load of 5% or more.

SELECTION OF ELECTRIC MOTORS

Each type of motor has its particular field of usefulness. Because of its simplicity, economy, and durability, the induction motor is more widely used for industrial purposes than any other type of ac motor, especially if a high-speed drive is desired.

If ac power is available, all drives requiring constant speed should use squirrel-cage induction or synchronous motors because of their ruggedness and lower cost. Drives requiring varying speeds, such as fans, blowers, or pumps, may be driven by wound-rotor induction motors. However, if there are machine tools or other machines requiring adjustable speed or a wide range of speed control, it will probably be desirable to install dc motors on such machines and supply them from the ac system by motor-generator sets or electronic rectifiers.

Almost all constant-speed machines may be driven by ac squirrel-cage motors because these motors are made with a variety of speed and torque characteristics. When large motors are required or when power supply is limited, the wound-rotor motor is used, even to drive constant-speed machines. A wound-rotor motor, with its controller and resistance, can develop full-load torque at starting with not more than full-load torque at starting, depending on the type of motor and the starter used.

For varying-speed service, wound-rotor motors with resistance control are used for fans, blowers, and other apparatus for continuous duty and are used for cranes, hoists, and other installations for intermittent duty. The controller and resistors must be properly chosen for the specific application. Synchronous motors may be used for almost any constant-speed drive requiring about 100 hp or over.

Cost is an important factor when more than one type of ac motor is applicable. The squirrel-cage motor is the least expensive ac motor of the three types considered and requires very little control equipment. The wound-rotor is more expensive and requires additional secondary control. The synchronous motor is even more expensive and requires a source of dc excitation, as well as special synchronizing control to apply the dc power at the correct instant. When very large machines are involved, as, for example, 1000 hp or over, the cost picture may change considerably and should be checked on an individual basis.

The various types of single-phase ac motors and universal motors are used very little in industrial applications, since polyphase ac or dc power is generally available. When such motors are used, they are usually built into the equipment by the machinery manufacturer, as in portable tools, office machinery, and other equipment. These motors are, as a rule, especially designed for the specific machine with which they are used.

chapter five

Manual Full-Voltage Motor Starters

A manual full-voltage motor starting switch is the simplest form of motor controller. On this type of controller, the contacts are operated by a direct mechanical linkage from a toggle handle or pushbutton to the contacts. The controlled motor is connected directly to the full value of circuit voltage for which the motor is rated.

One type of motor controller is shown in Fig. 5-1. This is a toggle-operated, single-pole switch with a thermal overload relay and is used on motor circuits where only one line needs to be interrupted (as on a 120-V motor of 1 hp or less). Another type of manual motor starter is shown in Fig. 5-2. This is a simple two-pole, fractional-horsepower toggle switch used for starting and stopping single-phase motors up to about 1 hp at 120 or 240 V. This is a basic snap-action switch that connects the motor to the line in the "on" position and disconnects it in the "off" position. The circuit shown in Fig. 5-2 is designed for a 240-V motor. Motor leads T1 and T3 connect directly to the switch terminals T1 and T3. A pilot light is connected in series with terminals T1 and T3, so that when the contacts are closed, causing current to flow to terminals T1 and T3, the lamp will light, indicating that current is flowing to the motor. This pilot light is optional and does not have to be used if not needed.

Manual starting switches for use with single-phase and polyphase motors of integral-horsepower ratings are used for motors rated up to approximately $7\frac{1}{2}$ hp, 600 V. In addition to the basic type, integral-horsepower manual starters are made in the reversing-type for manually reversing the direction of rotation of ac polyphase motors. Other types include controls for two-speed separately wound motors. Figure 5-3 shows a manual full-voltage motor controller for a three-phase motor.

Manual full-voltage starters are used on all kinds of machines and appliances, such as unit heaters, fans, pumps, small machine tools, and the like. Integral-horsepower manual starters are used for ac and dc motor control where remote

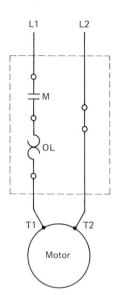

Figure 5-1 Single-pole manual motor control.

pushbutton control is not required or desired, where the operator is in attendance at the driven load and needs control there, and where conditions eliminate any hazard due to sudden restarting of motors upon return of power after a failure.

Three types of fractional horsepower motor starters are shown in Fig. 5-4. From left to right (a through c) is a single or one-pole starter, a two-pole starter, and a two-pole starter with a selector switch. In all three starters the ⓡ symbol indicates a thermal overload relay. Overload protection is provided in manual starters to open the circuit to the motor when excessive overload current, up to and including locked-rotor current (a condition in which the rotor is locked or held at a standstill), flows. This prevents the motor from operating at excessive current that would produce damaging heat in both the motor and supply circuit. Various overload devices that sense excessive current are used, including the solder-ratchet type and

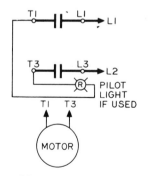

Figure 5-2 Two-pole manual motor control. (Courtesy Square D Company.)

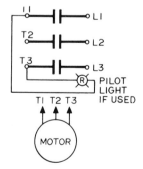

Figure 5-3 Manual motor starter for three-phase motor. (Courtesy Square D Company.)

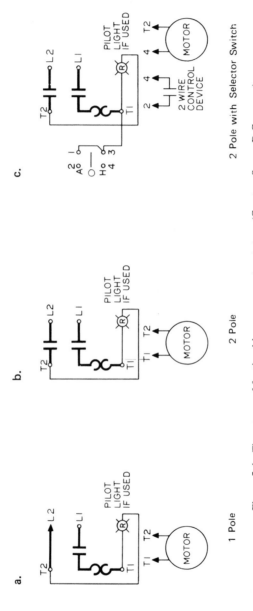

Figure 5-4 Three types of fractional-horsepower motor starters. (Courtesy Square D Company.)

a.

1 Pole

b.

2 Pole

c.

2 Pole with Selector Switch

the bimetallic type. Only one thermal overload relay is required in either the single-pole or double-pole controllers.

Solder-ratchet relay.   A small cylinder contains an alloy that will melt when an overload occurs. A small shaft to which a ratchet wheel is attached is embedded in the hardened alloy. When the contact mechanism of the starter is moved to the "on" position, a spring element holding the contacts closed engages the ratchet wheel. If an overload occurs, the heat produced by the excessive current flowing through a coil wrapped around the cylinder melts the alloy, releasing the ratched shaft, which can then no longer restrain the spring-loaded actuator held by the ratchet wheel. The actuator then moves to trip open the starter contacts. This type of relay is nontamperable and gives reliable overload protection. Repeated tripping does not cause deterioration and does not affect the accuracy of the trip point.

A wide variety of relay units is available so that the proper one may be selected on the basis of the actual motor full-load current. Since the motor circuit is actually in series with the overload relay, the motor will not operate unless the unit is complete with the relay installed.

Bimetallic relay.   A resistance heater wire is connected in series with the load current to the motor and is placed adjacent to a bimetallic element that will respond to heat from the heater when excessive current flows, causing the bimetal to deflect and open the contacts supplying the motor. Depending on the exact nature of the protective device, the starter can be reset for use after an overload by cooling and returning the handle to the "on" position.

REVERSING MANUAL STARTERS

Integral-horsepower manual starters are made in reversing type for manually reversing the direction of rotation of three-phase motors. One type of reversing manual motor starting switch is shown in Fig. 5-5. When the "forward" switch is executed

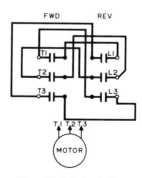

Figure 5-5   Reversing motor starter. (Courtesy Square D Company.)

Type K, 3 Pole, 3 Phase

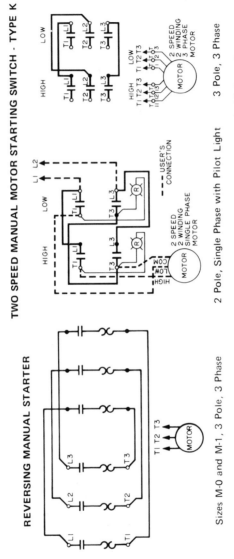

Figure 5-6 Starter similar to that shown in Fig. 5-5, except that circuits are protected with thermal overload relays. (Courtesy Square D Company.)

on this starter, the contacts on the left of the drawing close, causing current to flow to motor terminals T1, T2, and T3. However, if the "reverse" switch is pushed, the contacts on the right of the drawing close, which causes two of the line phases to change from their original position. This, in turn, causes the motor rotation to reverse.

The reversing manual starter in Fig. 5-6 is similar to the one in Fig. 5-5 except that the circuits are protected with thermal overload relays. Again, the set of contacts on the left of the drawing—when closed—will cause the motor to run in the forward motion. When the other contacts are manually closed, the forward contacts open, the reverse contacts close, causing two of the motor leads to be reversed, which in turn reverses the direction of the motor.

RULE: When any two of the three leads on a three-phase motor are reversed, the rotation of the motor will reverse.

Connection details of a three-phase, reversing, pushbutton-operated starter are shown in Fig. 5-7. The switches are mechanically interlocked so that the stop button must be depressed before directions are changed. The connections shown are for three-phase motors; for single-phase motors, the connection on lines L3 and T3 is omitted.

A two-speed pushbutton-operated starter with a thermal overload relay for each phase is shown in Fig. 5-8. Connector B is provided on motor starters that have two-coil overload protection. The mechanical interlock between starters allows speed transfer to be accomplished with the stop button being pushed. This starter is designed for two-speed, constant-horsepower, separate winding motors.

Diagrams of a three-pole switch with a mechanically linked pushbutton operator (without overload relay) are shown in Fig. 5-9. A double-pole switch with a mechanically linked pushbutton operator and thermal overload relay for single-phase motors is shown in Fig. 5-10. Both of these starters are of the full-voltage type.

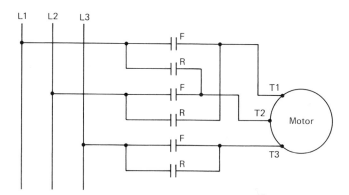

Figure 5-7 Connection details of a three-phase reversing starter.

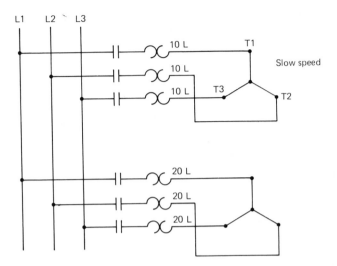

Figure 5-8   Two-speed pushbutton-operated motor starter.

Figure 5-11 shows the connection of a toggle-operated double-pole switch with thermal overload relay and with a ''run-Auto'' selector switch and a pilot light mounted in the same enclosure. The pilot light indicates when the motor is running. This type of circuit is often used with two-wire pilot devices such as thermostats.

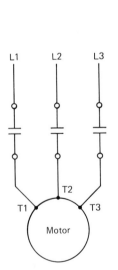

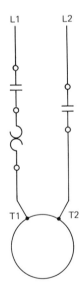

Figure 5-9   Diagram of a three-phase mechanically linked switch.

Figure 5-10   Diagram of a two-phase mechanically linked switch.

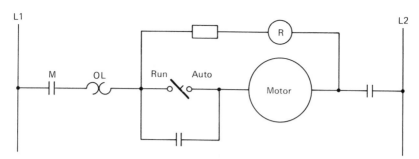

Figure 5-11   Toggle-operated two-phase switch with thermal overload relays.

DRUM SWITCHES

Drum switches are widely used for machine and other industrial applications. Such controllers contain switching elements in a vertical cylindrical housing that resembles a drum, from which it gets its name. These controllers are made in various types for both ac and dc motor control applications.

Manual reversing drum switches are used to start, stop, and reverse single-phase or inherently protected three-phase motors. The switch may be used for either maintained or momentary (spring return to center) contact operation. The manual drum switch is one of the most economical devices that can be installed, if the application does not require undervoltage or overload protection.

Three-phase type A forms of drum starters may be used on single-phase motors provided that one wire is connected directly to the motor. Single-phase type A forms may be used on three-phase motors provided that one wire is connected directly to the motor. In selecting a drum controller, first determine the type of motor to be controlled, the horsepower rating, and the type of mountings required. Then refer to manufacturers' catalogs for those available.

Drum switches often provide the most effective means of reversing or speed control of ac motors in cases where frequent switching is required and the controller is to be located close to the operator, such as on metal-turning lathes and other machine tools. Drum switches for reversing duty on three-phase motors simply provide reversal of two of the three lines' supply power to the motors.

SPECIAL NEC REQUIREMENTS

A switch used as a disconnecting means for a motor circuit must have a horsepower rating not less than that of the motor it controls. For example, a 5-hp motor must be started and stopped by a control device that has a nameplate rating of at least 5 hp. Certain exceptions to this rule are covered by Section 430 of the NEC.

If a magnetic switch is used as a motor controller, it must have a manually operated, disconnecting switch ahead of it. The switch may be a conventional safety switch, or in some cases, the overcurrent device (circuit breaker) may qualify as the disconnecting means. If the switch is more than 50 ft away from the motor, it must be in sight of the motor controller; otherwise, another switch must be provided in the circuit.

Switches used as disconnecting means for motor circuits should be rated in horsepower for all motors in excess of 2 hp. A switch should carry a rating of at least 125% of the full-load nameplate current rating of the motor and be manually operable in a readily accessible location. It must indicate whether it is in the open (off) or closed (on) position. When closed, the switch must disconnect both the controller and the motor from all ungrounded supply conductors.

SUMMARY

Fractional-horsepower manual starters are used whenever it is desired to have overload protection as well as on–off control of small ac single-phase or dc motors. The NEC requires overload protection for fractional-horsepower motors whenever they are started automatically or by remote control. The motor is connected directly across the line on start, which is usually not objectionable with motors of 1 hp or less.

The chief disadvantage of manual starters is that low-voltage protection and low-voltage release are not possible. Therefore, if there is a power failure, the contacts will remain closed and the handle will stay in the "on" position. When power is restored, the motor will automatically restart. Although this is a disadvantage for some applications, in others—such as fans, blowers, pumps, and oil burners, where the motors should run continuously—this feature could be an advantage.

The compact physical size of these manually operated controls permits mounting directly on the machinery and in various other places that would not be suited for the larger magnetic-type starters. In fact, the unenclosed device can be mounted in a standard switch or conduit box and can be covered with a standard flush plate. The on–off positions are clearly marked on the operating lever, which is very similar to that used in a standard toggle switch used for lighting control.

Fractional-horsepower manual starters are available in several different types of enclosures, as well as the open type. Enclosures can be obtained to shield the live parts from accidental contact, to mount in machine cavities, to protect the starter from dust and moisture, or to prevent the possibility of an explosion when the starter is used in hazardous locations.

Besides the applications mentioned previously in this chapter, automatic control devices such as pressure switches, float switches, or thermostats may be used in conjunction with fractional-horsepower manual starters. Their contact capacity must, however, be sufficient to make and break the full motor current.

Integral-horsepower manual starters may be used where it is convenient for the operator to start and stop small single-phase or three-phase motors by pressing pushbuttons mounted in the cover of the starter enclosure. They may be used satisfactorily when the full motor starting torque may be safely applied to the driven machine and when the current inrush resulting from application of line voltage is not objectionable.

Again, as with fractional-horsepower starters, low-voltage protection and low-voltage release are not obtainable with the manually operated mechanism. If the power fails, the starter contacts will remain closed until the stop button is manually operated. Therefore, when there is a power outage, the motor will restart when the power is restored, unless, of course, the operator has opened the circuit in the meantime. However, these are very popular controls where applicable because the initial cost is low, motor overload protection is provided, and operation is both safe and economical.

Some classes of integral-horsepower manual starters are provided with thermal overload protection, including "trip-free" operation. If the overload relays act to open the starter contacts, the contacts cannot again be closed until the relays have been reset by pressing the stop button. This arrangement prevents the operator from holding the motor circuit closed by the start button when the overload relays have operated to protect the motor. If the relays are reset by operating the stop button and the starter is again closed while the motor is still overloaded, the relays will continue to trip until the overload is removed.

Usual applications are the control of small machine tools, punch presses, fans and blowers, grinders and buffers, and overload protection for pumps, compressors, and portable electric tools. Automatic control devices may also be used with these starters if their contact capacity is sufficient to make and break the motor line current.

MAINTENANCE OF MANUAL STARTERS

On manual starters that employ a toggle switch with a quick-make and quick-break, there is practically no maintenance except for checking tightness of connections and being sure that heaters for overload relay are tight.

On the oil-immersed drum type, it is well to:

1. Check all connections.
2. Observe wear on removable contact tips and replace when two-thirds worn away.
3. Replace oil when it becomes dirty or heavily carbonized.
4. See that all parts are clean and move freely.

chapter six

AC Magnetic Starters

Magnetic starters utilize contacts operated by an electromagnetic coil. The contacts close when the coil is energized and open when the coil is deenergized. The coil circuit is energized by a switching device that makes and breaks the circuit to the coil.

Magnetic starters of the full-voltage type apply the full-rated voltage to the motor when the contacts are closed. In addition to the contacts and the electromagnetic operating coil, magnetic starters include running overload protective devices that will disconnect the motor from its source of supply when overloaded, up to and including a locked-rotor condition. Across-the-line magnetic starters are made for both single-phase and polyphase motors.

The operating coil of the magnetic starter is a single-phase two-wire load device commonly connected across two of the supply wires to the motor. A switching device connected in series with the coil provides on–off control of current flow through the coil. This device is commonly a set of pushbuttons (start and stop) mounted in the front of the starter enclosure or at a location away from the starter to provide remote control of the motor.

Square D Class 8501 magnetic relays are used as auxiliary devices for switching control circuits and for controlling small motors or other light loads, such as electric heaters, pilot lights, or audible signals. Class 8502 contactors are available in sizes from 0 through 6 and provide a safe and convenient means for connecting and interrupting of branch circuits. These contactors are usually used on electric motor loads where overload protection is not required or is provided separately. Pilot devices such as pushbuttons, float switches, pressure switches, limit switches, or thermostats are used to provide the necessary control for operating these contactors.

Square D Class 8536 line-voltage-type magnetic starters provide a convenient and economic means for starting and stopping ac squirrel-cage motors. They are generally used where a full-voltage starting torque may be safely applied to the driven machinery and where the current inrush resulting from across-the-line starting is not objectionable. These starters are usually controlled by pilot devices such as pushbuttons, limit switches, or timing relays.

Magnetic motor starters are available as two-, three-, or four-pole contactors with one, two, or three overload relays for both single- and three-phase applications, with full-voltage nonreversing ac single-speed motors up to 200 hp, 600 V maximum. They are also available for use with small nonreversing dc motors up to and including $1\frac{1}{2}$ hp at 120 V or 2 hp at 240 V.

General Electric starters are available in NEMA sizes 00 through 5, in open forms, general-purpose enclosures, water- and dust-tight enclosures, and hazardous-duty types. A flush-mounted enclosure for use in plaster walls or machine tool cavities is also available.

In selecting this type of starter consideration should be given to the horsepower rating of the motor, whether single- or three-phase, the coil voltage, the control power transformer, the type of enclosure, and the overload relay heaters.

Magnetic reversing across-the-line controllers are frequently used for starting of single-phase or three-phase motors up to 200 hp, 600 V maximum where the application requires a reversing or plugging action.

When the voltage on the coil of an ac contactor or relay passes through zero its magnetic pull or holding power is zero and the device starts to open. The voltage, however, is soon effective in the opposite direction and the device is again pulled closed. This operation causes a humming noise in any ac-operated device and a decided chattering noise in a defective unit. The otherwise objectionable chattering is eliminated and the device is kept closed by the use of a shading coil, usually embedded in the laminated magnetic circuit of the device. The shading coil produces enough out-of-phase flux to provide holding power to maintain the device closed during the short period when the power to maintain flux is zero. Even with shading coils in use, the air-gap surfaces must be free from dirt and well fitted to avoid objectionable noise. Broken shading coils are ineffective and cause noisy operation.

For quiet operation of ac contactors, it is necessary to provide well-fitted surfaces at the air gap. Any dirt in this area introduces a greater air gap when the unit is closed, increases the duty imposed on the shading coil, and results in noisier operation. To prevent rusting of the fitted surfaces at the air gap, these devices are often shipped with a small amount of grease or oil on them. This lubrication may cause a ''seal'' that makes them sticky and sluggish in opening when first put into service. These surfaces should be wiped clean before the units are placed in service.

Dc coils are not subject to a zero-voltage condition. Therefore, dc-operated devices are always quiet. For this reason, ac current-carrying contactors equipped with dc operating coils will operate quietly.

MAGNETIC STICKING

When operating coils are deenergized, some residual magnetism remains in the magnetic circuit. This residual magnetism is sometimes strong enough to hold the device closed after the coil is deenergized. This condition occurs most frequently on small devices on which contact spring pressures and moving parts are light. Magnetic sticking causes erratic, unsatisfactory, and sometimes dangerous operation. It can be avoided by adding a nonmagnetic shim in the magnetic circuit.

MAINTENANCE OF MAGNETIC STARTERS

1. Do not lubricate contact tips or bearings.
2. Wipe magnet sealing surfaces occasionally with an oil-moistened cloth to prevent noise and rust.
3. Check tightness of all connections, especially connections to overload heaters, since a loose connection here will cause local heating that will affect the calibration of the relay.
4. Make sure that shunts are not broken or touching other parts.
5. Adjust contacts so that they all meet at the same time.
6. In general, the contacts will not need attention during normal life. If they become excessively rough or burned in service, dress them with a fine file. Do not use emery cloth. Replace contact tips when approximately two-thirds of their thickness is worn away. These are movable, and only a screwdriver is needed for the change.
7. Remove any excess deposits from the inside surfaces of the arc boxes adjacent to the contacts, and replace any broken arc boxes.
8. See that all moving parts work freely.
9. Disconnect the motor and manually test the start button, the stop button, the overload relay, and the reset.
10. Most industrial linestarters are provided with overload relays whose action depends on the movement of a bimetallic strip under heat. On very small motors the bimetallic strip actually carries the motor current, but on larger motors a separate heater carrying the motor current is placed close to the strip. On still larger motors, a current transformer reduces the motor current to a value that can be handled by the heater.

THERMAL OVERLOAD RELAYS

Thermal overload relays have the inverse-time-limit feature, which means the greater the overload, the shorter the time of tripping. They provide excellent protection against overloads and momentary surges but do not protect against short-circuit currents. For protection against the latter, fuses not exceeding four times the motor full-load current, time-limit circuit breakers set at not more than four times the motor full-load current, or instantaneous trip circuit breakers should be installed ahead of the linestarter. Where fuses are used it is advisable to use a disconnecting switch as well.

Heaters for thermal relays are made with different current ratings, so that within its limits any starter can be used with different sized motors and still afford proper protection by selecting the size of heater that corresponds to the full-load current of the motor being used. In general, the ampere rating of a heater should be approximately 120% of the motor full-load current.

A calibration lever on some types of relays makes additional adjustment possible. When set at 100%, the current stamped on the heater will just trip the starter after several minutes. For tripping at a smaller current, the lever is moved toward 90; at a larger current, the lever is moved toward 120.

MAGNETISM

Anyone working in the field of electric motor controls should be familiar with the principles of magnetism, because electric motors and many types of motor starters depend on magnets and magnetism for their operation. Therefore, a review of magnetism follows.

A magnet is either permanent or temporary. If a piece of iron or steel is magnetized and retains its magnetism, it is a permanent magnet. A compass is one form of permanent magnet. Others with which most readers are probably familiar are horseshoe-shaped magnets and bar magnets. Each of these magnets has a north magnetic pole and a south magnetic pole, as do all magnets.

When current flows through a coil, a magnetic field with a north and a south pole is set up just like that of a permanent magnet. However, when the current stops, the magnetic field also disappears. This temporary magnetism is called *electromagnetism*. Permanent magnets are used for the magnetic field necessary in the operation of small, inexpensive electrical motors. Larger motors, relays, and transformers rely on the magnetic fields from electrical current passing through coils of wire.

When electricity flows through a wire or conductor, magnetic lines of force (magnetic flux) are created around that wire (Fig. 6-1). When a piece of wire is passed through a magnetic field (magnetic lines of force), electricity is created in

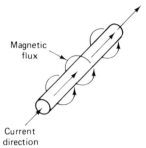

Magnetic
flux

Current
direction

Figure 6-1   When electricity flows through a conductor, magnetic lines of force are created around the conductor.

that wire. We can then readily see the relation between electricity and magnetism. In fact, the very existence of the electrical industry is dependent on magnetism and magnetic circuits. If the current in a conductor is flowing from south to north and a compass is placed under the conductor, the north end of the needle will be deflected to the west; if the compass is placed over the conductor, the north end of the needle will be deflected to the east (Fig. 6-2). Here are four basic rules to follow when working with magnetism:

1. *To determine the polarity of an electromagnetic solenoid:* In looking at the end of a solenoid, if an electric current flows in it clockwise, the end next to the observer is a south pole and the other end is a north pole; if the current flows counterclockwise, the position of the poles is reversed.

2. *To determine the direction of the lines of force set up around a conductor:* If the current in a conductor is flowing away from the observer, the direction of the lines of force will be clockwise around the conductor.

3. *To determine the direction of motion of a conductor carrying a current when placed in a magnetic field:* Place the thumb, forefinger, and middle finger of the left hand at right angles to each other; if the forefinger shows the direction

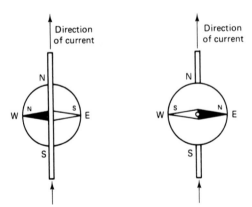

Figure 6-2   Direction of compass needle in relation to current flow.

of the lines of force and the middle finger shows the direction of the current, the thumb will show the direction of the motion given to the conductor.

4. *To determine the direction of an induced current in a conductor that is moving in a magnetic field:* Place thumb, forefinger, and middle finger of the right hand at right angles to each other; if the forefinger shows the direction of the lines of force and the thumb shows the direction of motion of the conductor, the middle finger will show the direction of the induced current.

A rule that is sometimes useful is the following: If the effect of the movement of a closed coil is to diminish the number of lines of force that pass through it when viewed by a person looking along the magnetic field in the direction of the lines of force—but if the effect is to increase the number of lines of force that pass through the coil—the current will flow in the opposite direction.

chapter seven

AC Magnetic Reversing
Controllers and Combination
Starters

Across-the-line magnetic reversing motor starters are used for full-voltage frequent starting of single-phase of polyphase motors up to about 200 hp, 600 V maximum, where the application requires reversing or plugging operation. Three-phase squirrel-cage motors are particularly suited to reversal of rotation since reversing the connection of any two of the motor feeds will cause a change in rotation. This reversal of line connection is normally done by using two separate contactor assemblies—one for the forward rotation and the other for reverse rotation. Both contactor assemblies are mounted in a single enclosure or cabinet. Reversing starters are used for starting, stopping, and reversing of three-phase squirrel-cage motors, for primary reversing of wound-rotor motors, and for some single-phase applications. These starters are available with or without running overload protection, and they are also available in combination-starter form, that is, with an on–off disconnect switch—either fusible or nonfusible.

Open reversing controllers can also be furnished with either horizontal or vertical mechanical interlocking. Momentary-contact pushbutton control devices provide complete undervoltage protection for the load circuit. Thermal overload relays provide complete overcurrent protection.

Typical control circuits for AC reversing magnetic starters are shown in Figs. 7-1 and 7-2. Another reversing circuit is shown in Fig. 7-3. Note that the contacts (F) of the forward contactor, when closed, connect lines 1, 2, and 3 to motor terminals T1, T2, and T3, respectively. As long as the forward contacts are closed, mechanical and electrical interlocks prevent the reverse contactor from being energized.

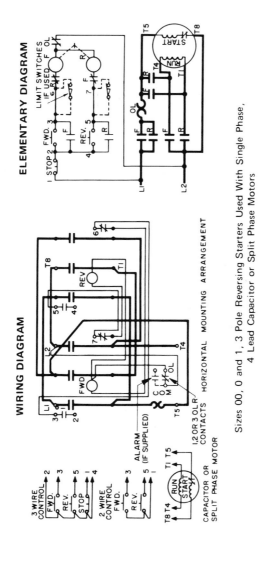

Figure 7-1 Typical wiring diagrams of ac reversing magnetic starters. (Courtesy Square D Company.)

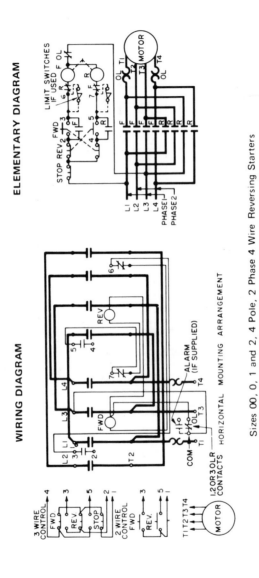

ELEMENTARY DIAGRAM

WIRING DIAGRAM

Sizes 00, 0, 1 and 2, 4 Pole, 2 Phase 4 Wire Reversing Starters

Figure 7-1 (continued)

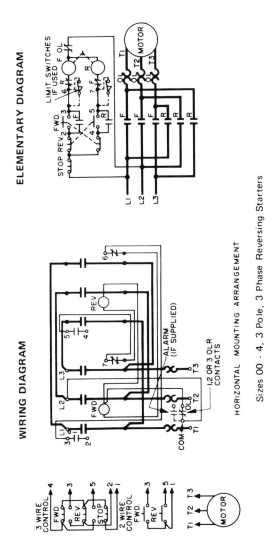

ELEMENTARY DIAGRAM

WIRING DIAGRAM

HORIZONTAL MOUNTING ARRANGEMENT

Sizes 00 - 4, 3 Pole, 3 Phase Reversing Starters

Figure 7-1 (continued)

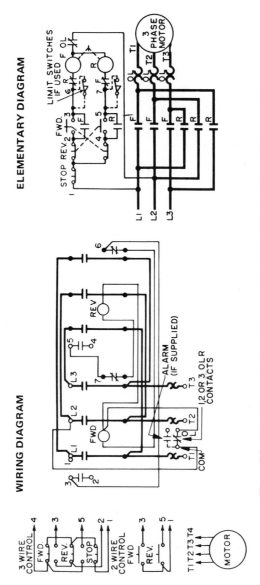

Figure 7-2 Size 5 three-pole three-phase reversing starter. (Courtesy Square D Company.)

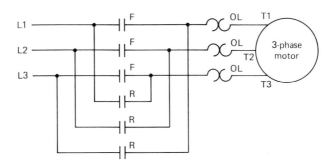

Figure 7-3   Typical reversing motor starter.

When the forward contactor is deenergized, the second contactor can be picked up, closing its contacts (R), which reconnect the lines to the motor. Note that by running through the reverse contacts, line 1 is connected to the motor T3, and line 3 is connected to motor terminal T1. The motor will now run in the opposite direction.

Whether operating through either the forward or reverse contactor, the power connections are run through an overload relay assembly, which provides motor overload protection. A magnetic reversing starter therefore consists of a starter and contactor, suitably interwired, with electrical and mechanical interlocking to prevent the coil of both units from being energized at the same time.

Manual reversing starters (employing two manual starters) are also available. As in the magnetic version, the forward and reverse switching mechanisms are mechanically interlocked, but since coils are not used in the manually operated equipment, electrical interlocks are not furnished.

AC COMBINATION STARTERS

With minor exceptions, the NEC requires a disconnect means for every motor. A combination starter consists of a magnetic starter and a disconnect means, all in a common enclosure. Most starters of this type include a visible blade disconnect switch, either fusible or nonfusible; other types employ a circuit breaker.

A combination starter provides greater safety for the operator because the cover of the enclosing case is interlocked with the external operating handle of the disconnect means. The door cannot be opened with the disconnect means closed. With the disconnect means open, access to all parts may be had, but no hazard is involved because there are no parts connected to the power line. This safety feature cannot be obtained with separately enclosed devices. In addition, the cabinet is provided with a means for padlocking the disconnect in the "off" position.

Line-voltage combination starters are used where full-voltage starting torque may be safely applied to the driven machine, and where the resulting current inrush

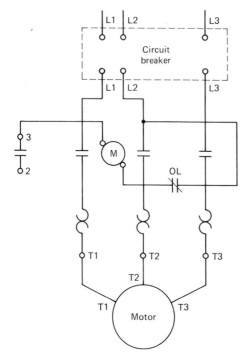

Figure 7-4   Typical schematic diagram of a combination magnetic motor starter.

is not objectionable. If starting torque must be reduced, or current inrush limited, a reduced voltage starter should be used. Line-voltage combination starters may also be used as primary switches for slip-ring (wound-rotor) motors, when manual speed regulators are provided for the secondary circuit.

Combination nonreversing magnetic starters are used for full-voltage starting of single-speed squirrel-cage induction motors up to 200 hp, 600 V maximum. They also provide motor overcurrent protection with bimetallic overload relays. Under-voltage can be obtained by using a momentary-contact pushbutton. Combination starters are suitable for remote control with a pushbutton station, manual control switch, or automatic control accessories.

When selecting magnetic combination starters, consider the following:

1. Horsepower of motor to be controlled
2. Fuse-clip size or circuit breaker
3. Coil voltage and/or control power transformer
4. Enclosure type
5. Modifications
6. Overload relay heater selection

A wiring diagram of a combination motor starter is shown in Fig. 7-4.

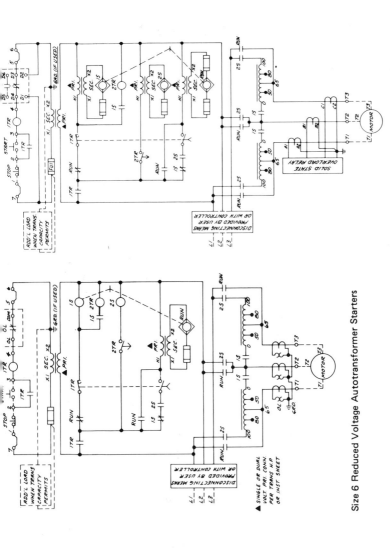

Figure 8-1 Typical wiring diagram of an autotransformer motor starter. (Courtesy Square D Company.)

Size 6 Reduced Voltage Autotransformer Starters

Size 7 Reduced Voltage Autotransformer Starters

chapter eight

AC Reduced-Voltage Motor Starters

Any three-phase motor can be connected directly to the full voltage for
rated without doing damage to the motor. But the inrush of starting curr
voltage—which can be as much as eight (or more) times the norma
current—can cause severe voltage disturbances in the electrical system fi
the motor is supplied. In addition, the shock of the starting torque of
motors might damage the driven load in some cases. As a result, it may
sary to start a motor at reduced voltage to prevent or minimize any obje
effects on the load or the supply system. Starters used for such cases
reduced-voltage starters.

Reduced-voltage starters, both manual and automatic, have a broad
application. Reduced-voltage starters of the autotransformer type find
application where the size or design of the motor, or restrictions of th
circuit, requires starting on reduced voltage. The autotransformer-type st
vides greater starting torque per ampere starting current drawn from the
any other type of reduced-voltage motor starter.

One way to start a motor at reduced voltage is to connect a resistor
conductor to the motor. The resistor limits the current and reduces the
applied to the motor windings. There are both manual- and automatic (ma
type resistance starters that first connect resistance in the motor circuit to re
voltage to the motor and hold down starting current. Once the motor re
certain speed, the resistance is cut out of the circuit and full voltage is applie
motor. In manual units, a handle is moved from the start to the run position
the motor up to speed and normal operation. Magnetic contactors cut
resistance in the automatic resistance starters.

Probably the most widely used type of reduced-voltage starter today is the autotransformer or compensator starter. This starter limits the starting current with higher starting torque than other reduced-voltage starters without the energy loss of resistor starters. As with resistance starters, autotransformer starters are available in both manual and automatic form. In the manual starter, a handle is first moved to the start position, held a few seconds, and then moved to the run position. In the start position, tap connections are made on an autotransformer assembly in the unit to apply to the motor a voltage less than the full circuit voltage. When the motor has come up to a certain speed, the autotransformer winding is cut out and full voltage is applied to the motor. In the automatic autotransformer starter, the switching operations from reduced voltage to full voltage are made by magnetic contactors operated by timing devices.

Autotransformer starters are normally rated from 5 to 250 hp. Two taps are provided for 65 and 80% of line voltage in most cases. However, for use on motors above 50 hp, taps are provided for 50, 65, and 80% of the line voltage, giving respective line currents equal to 25, 42, and 64% of the full-voltage starting current. A typical wiring diagram of an autotransformer is shown in Fig. 8-1.

When selecting an autotransformer starter, the following information should be researched:

1. Motor voltage
2. Startup torque required

   Caution: The transition from start to run is open (momentary loss of current to motor) with the manual autotransformer starter. If closed transition is required (power is continuously applied from start to run), use a magnetic form of starter.

3. Heater selection
4. Enclosure type: NEMA 1 or NEMA 3R
5. Shallow mount or with conduit access box
6. Modifications, including combination forms

A magnetic reduced-voltage starter circuit is shown in Fig. 8-2. This starter features closed-circuit transition with no interruption in line current during the transition. A pneumatic timing relay permits easy adjustment of starting-time period on reduced voltage.

Common uses for this starter include starting duty for blowers, conveyors, and pump motors in connection with automatic pilot devices such as limit switches and pressure switches.

When selecting magnetic reduced-voltage starters, the following should be given consideration:

1. Operation duty cycle
2. Voltage and type of motor to be controlled

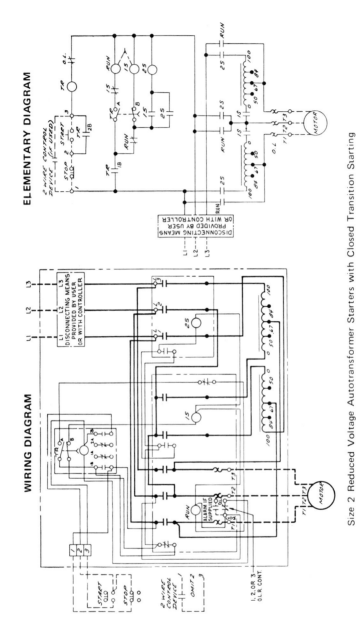

Size 2 Reduced Voltage Autotransformer Starters with Closed Transition Starting

Figure 8-2 Magnetic reduced-voltage starter circuit. (Courtesy Square D Company.)

ELEMENTARY DIAGRAM

WIRING DIAGRAM

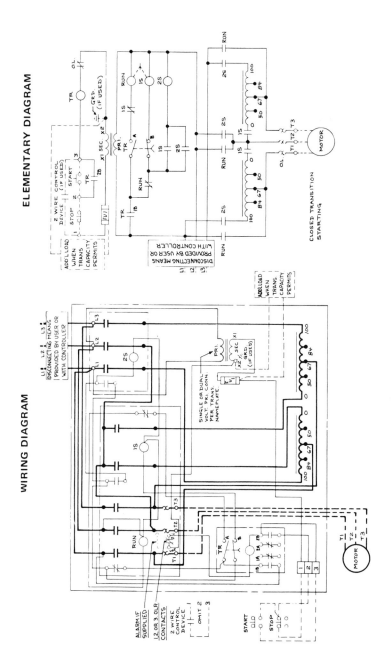

Sizes 3 and 4 Reduced Voltage Autotransformer Starters with Closed Transition Starting, Control Circuit Transformer and Secondary Fuse (Form FT)

Figure 8-2   (continued)

141

3. Overload relay heater selection
4. Type of enclosure
5. Modifications
6. Combination form required

PART-WINDING STARTERS

Part-winding magnetic starters (Fig. 8-3) are sometimes referred to as *increment starters*. These starters are commonly used to control motors driving light or low-inertia loads, such as air-conditioning compressors, refrigeration compressors, pumps, fans, and blowers. This method of starting has its limitations on the type of load that can be accelerated on the first point. Inrush current is limited to an average of approximately 65% of across-the-line starting current, depending on the use of either two-thirds or one-half of the motor winding on start. Part-winding magnetic starters are available in NEMA sizes 1PW to 5PW, types 1, 4, and 12, as well as open forms.

When selecting this type of motor, the following should be given consideration:

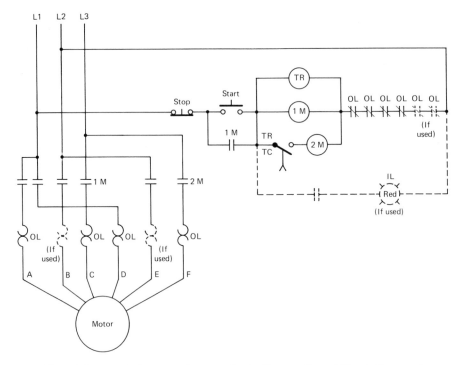

Figure 8-3   Part-winding magnetic starter circuit. (Courtesy General Electric Company.)

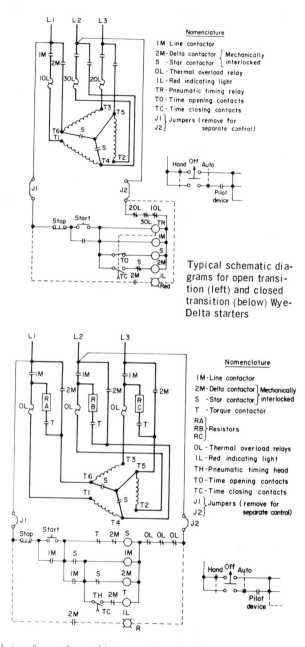

Typical schematic diagrams for open transition (left) and closed transition (below) Wye-Delta starters

Figure 8-4 Various forms of wye–delta magnetic starters. (Courtesy General Electric Company.)

1. Type of motor to be controlled (must be reconnectable, separate, isolatable windings suitable for use with part-winding starter)
2. Startup torque requirements
3. Heater selection
4. Enclosure
5. Modifications

WYE–DELTA MAGNETIC STARTERS

Wye–delta magnetic starters are for use with low-starting-torque applications such as fans, compressors, and conveyors driven by motors capable of being connected in wye and in delta. Wye–delta starting provides a low inrush current which results in low starting torque. When the motor windings are connected in wye during starting, each winding has 58% full voltage. Automatically reconnecting to delta on run applies full voltage to each winding.

Either open or closed transition forms are available in open baseplate forms in general-purpose enclosures and in watertight or industrial-use enclosures (see Fig. 8-4).

When selecting this type of starter, determine the following:

1. Type of motor to be controlled (must be nameplate rated for starting with wye–delta starters)
2. Startup torque required
3. Voltage and horsepower of motor
4. Combination form required
5. Overload relay heater selection
6. Modifications

MAGNETIC PRIMARY RESISTOR STARTERS

Resistance-type starters (Fig. 8-5) are sometimes applied on network distribution systems where power companies' regulations require that the circuit not be opened during the transition from reduced voltage to full voltage. They are especially desirable where sudden mechanical shock to the driven load must be avoided.

Automatic reduced-voltage starters are well adapted for geared or belted drives, where a sudden application of full-voltage torque must be avoided. The inrush current is limited to approximately 80% of the full-voltage locked-rotor value and approximately 64% of the full-voltage locked-rotor starting torque is produced.

General-start-duty NEMA class 116 resistors or high-inertia-start-duty NEMA class 156 resistors are both available in this model of motor starter.

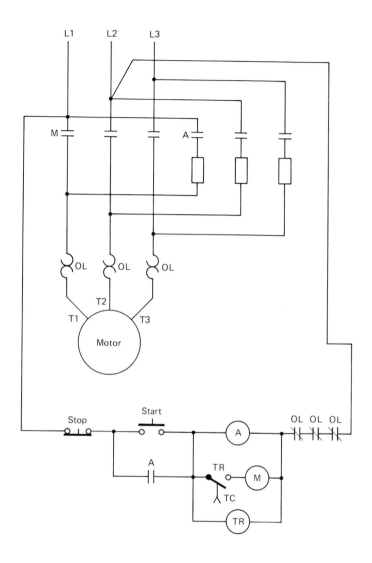

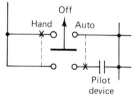

Figure 8-5 Resistance-type starter. (Courtesy General Electric Company.)

TYPE OF STARTER	STARTING CHARACTERISTICS IN PER CENT			ADVANTAGES	LIMITATIONS
	Voltage at Motor	Line Current	Torque		
FULL VOLTAGE	100	100	100	1. Lowest cost. 2. Least Maintenance. 3. Highest starting torque.	1. Starting current may exceed limits of electrical distribution system. 2. Starting torque may be too high for some machines.
AUTOTRANSFORMER (manual or magnetic)	80 65 50	64 42 25	64 42 25	1. Provides maximum torque per line ampere for reduced-voltage starters (high torque efficiency). 2. Starting characteristics easily adjusted. 3. Suitable for wide variety of applications. 4. Closed circuit transition (magnetic only).	1. Most complex of reduced-voltage starters because proper sequencing of energization must be maintained. 2. Duty cycle (frequency and length, in time, of starting) may be limited by standard autotransformer rating.
PRIMARY RESISTOR	80	80	64	1. Least complex method to obtain reduced-voltage starting characteristics on low voltage systems because interlocking of contactors is unnecessary. 2. Smoother acceleration. 3. Up to 6 points of acceleration. 4. Improves starting power factor because voltage-current lag is shorter by putting a resistance in series with the motor. 5. Less expensive than autotransformer type through size 3. 6. Closed circuit transition.	1. Additional power loss in resistors compared to other types of starters. 2. Low torque efficiency (decreases as voltage is decreased). 3. Starting characteristics not easily adjusted after manufacture. 4. Duty cycle may be limited by standard resistor rating.
PART-WINDING (2/3 winding)	100	65	42	1. Suitable for low or high voltage. 2. Closed circuit transition. 3. Full acceleration in one step for most standard induction or special part-winding motors when 2/3 winding connection is used.	1. Torque efficiency usually poor for high speed motors. 2. Possibility of motor not fully accelerating due to torque dips. 3. Usually not suitable for high inertia loads. 4. Specific motor types required.
WYE-DELTA	100	33	33	1. Starting duty cycle not limited by control (motor heating only limit). 2. High torque efficiency. 3. No torque dips or unusual winding stresses occur as with part-winding starting.	1. Starting characteristics not adjustable (depend upon motor design). 2. Requires motor capable of wye and delta connection. 3. Control more complex than other starter types.

Figure 8-6 Reduced-voltage starter comparison table. (Courtesy General Electric Company.)

The following should be considered when making selection of starters:

1. Start duty (torque requirement)
2. Operational voltage and frequency
3. Horsepower of motor
4. Combination form required
5. Overload relay heater selection
6. Enclosure selection

The reduced-voltage starter comparison table in Fig. 8-6 should prove useful in selecting reduced-voltage starters.

MAINTENANCE OF REDUCED-VOLTAGE STARTERS

In general, the maintenance and troubleshooting points covered in Chapter 15 apply to reduced-voltage starters. In addition, the following points should be noted:

1. Do not lubricate contact tips or bearings.
2. Wipe magnet sealing surfaces occasionally with an oil-moistened cloth to prevent noise and rust.
3. Check tightness of all connections, particularly connections to overload heaters, since a loose connection here will cause local heating that will affect the calibration of the relay.
4. Make sure that shunts are not broken or touching other parts.
5. Adjust contacts so that they will all meet at the same time.
6. In general, the contacts will not need attention during normal life. If they become excessively rough or burned in service, dress them with a fine file. Do not use emery cloth. Replace contact tips when approximately two-thirds of their thickness is worn away. These are removable, and only a screwdriver is needed for the change.
7. Remove any excess deposits from the inside surfaces of the arc boxes adjacent to the contacts, and replace any broken arc boxes.
8. See that all moving parts work freely.
9. Disconnect the motor and manually test the start button or handle.
10. Many of these starters previously used an overload relay of the magnetic type having an oil dashpot to prevent tripping on momentary overloads. Take care to see that the piston is not binding in the dashpot and that the dashpot is filled to cover the piston with the proper grade of dashpot oil.
11. Check the setting of the overload relay and be sure that there is at least 125% full-load motor current. Read the setting where the calibration line on the dashpot coincides with the bottom of projection on the upper casting.

TABLE 8-1 NEMA Classification of Resistors[a]

Percent of Full-Load Current on First Point	Starting Torque[b] (% of Full Load)					Resistor Class Number						
	Series Motors	Compound Motors	Shunt Motors	Wound-Rotor Induction Motors		5 Sec on Out of 80 Sec	10 Sec on Out of 80 Sec	15 Sec on Out of 90 Sec	15 Sec on Out of 60 Sec	15 Sec on Out of 45 Sec	15 Sec on Out of 30 Sec	Continuous
				One-Phase Starting[c]	Three-Phase Starting[c]							
25	8	12	25	15	25	111	131	141	151	161	171	91
30	30	40	50	30	50	112	132	142	152	162	172	92
70	50	60	70	40	70	113	135	143	153	163	173	93
100	100	100	100	55	100	114	134	144	154	164	174	94
150	170	160	150	85	150	115	135	145	155	165	175	95
200	250	230	200	—	200	116	136	146	156	166	176	96

[a]The duty cycles in this tabulation are more nearly in accord with actual practice than those in the older AES classification.

[b]Based on Westinghouse motors.

[c]This refers to the connections of the rotor circuit.

Source: Courtesy Westinghouse.

12. Above 35 hp at 25 Hz and above 50 hp at 60 Hz, and above 550 V, it was customary at one time to use insulating oil in the pan so that contacts would make and break under oil. Today, autostarters use thermal relays and no oil pans are necessary. When used, however, the pan should be filled with oil to the oil-level mark. If the oil becomes badly discolored or carbonized from service, replace it with new oil after carefully cleaning the pan.

13. In automatic starters of this type, a definite time relay is used to control the time on the starting tap. The timing of this relay should be checked, and if a type using an oil dashpot is employed, care must be taken to see that it is filled with oil. This is a very light special oil, with a low pour test, suitable for low temperatures.

14. The starting transformers are provided with taps by which the starting voltage can be varied. The proper taps to use are those that will bring the motor up to speed in 20 seconds or less on motors up to 100 hp. On 220-hp motors the starting period may be 30 seconds. Periods longer than these may seriously overheat the transformer.

Since almost all starters of the resistance type employ drum-type controllers in the smaller sizes, and contactors in the larger, the maintenance requirements are practically the same as covered for other types of starters. This type of starter is called resistance type, because resistance inserted in the primary circuit is cut out gradually by the starter as the motor comes up to speed. It is usually designed to short-circuit the resistor in from 1 to 10 seconds and for starting not more than once every 80 seconds.

If the resistors are properly applied and if the connections are maintained tight, no further service should be required. Investigate any excessive heating of the resistor to determine if it was caused by open or unbalanced connections in the secondary circuit. On new installations, duty cycle and service should be checked against the resistor class to verify proper application, since all have definite application and limitations. For example, NEMA class 114 is for starting duty only and on the basis that the motor should be started and brought up to speed in approximately 5 seconds with a minimum of 75 seconds between successive starts.

NEMA classifications of resistors are shown in Table 8-1.

chapter nine

Miscellaneous Motor Controls

Besides the conventional starters discussed previously in this book, many control circuits utilize other control devices that are used in conjunction with motor starters for a more refined control of the motors in a particular application.

PNEUMATIC AND MOTOR-DRIVEN TIMING RELAYS

Pneumatic timing relays are used in automation circuits, such as machine-tool sequencing operations, process-industry operations, conveyor lines, and other applications where a time-delay device is required. The typical relay may be adjusted for a time delay of between $\frac{1}{5}$ and 18 seconds by the turning of a single screw, which is easily accessible from the front of the relay.

Most relays of this type can be obtained either as a time delay after energizing the coil or after deenergizing the coil, or it can be converted in the field from one form to the other.

Motor-driven timing relays are for use in ac circuits where a time delay or accurate timing of operation is required. Typical uses of this type of relay are in sequencing multiple operations, performing definite time operations or in connection with a reduced-voltage starter (Chapter 8) for accelerating large motors. Various forms are available for time delays from 3 seconds to as much as 400 minutes. Settings are made by turning a large insulated knob until the adjustment knob indicates the desired time. The adjustment knob also gives an indication of the unexpired time during the entire time cycle.

In selecting a time-delay relay, consideration must be given to the following:

1. Time delay on energization only
2. Time range required
3. Type of contacts
4. Coil voltage
5. Enclosure

Solid-state timers are suitable in applications where high repeat accuracy is desired and where time delay must be frequently changed.

Proper selection of timing relays of this type for a particular application can be made only after a study of the service requirements and with a knowledge of the operating characteristics inherent in each available device. However, the construction and performance features of the pneumatic timer make it suitable for the majority of industrial applications.

The Square D Class 9050 Type X Digital Timing Relay offers great flexibility, high accuracy, and ease of maintenance, all at a reasonable cost. Each timer has a built-in flexibility of seven timing ranges (0.01 second to 999 hours), any of which can be conveniently selected by switching an internal rocker switch. The type X timer offers a repeat accuracy of ± 10 milliseconds (based on a constant frequency) in any timing range. The output contacts consist of two miniature plug-in relays, allowing easy replacement should the need arise.

Three types of relays are shown in Fig. 9-1.

CONTROL TRANSFORMERS

Control circuit transformers (Fig. 9-2) are specially designed for industrial control applications to provide good transformer regulation when high inrush currents are drawn. Their principal use involves reducing control circuit voltages for many reasons, the most important of which are as follows:

1. Operator safety is increased by the use of low voltage at the control station and other pilot devices. Also, because of the reduced potential, there is less chance of a fault occurring between lines of the control circuit wiring or to ground.
2. Simplifies voltage changeover of complicated control panels—the transformer is replaced, or, in the case of a transformer with a dual primary, the primary windings are reconnected instead of changing the magnet coil of each device.

NEMA standards require magnetic devices such as starters to operate satisfactorily at 85% of rated voltage. Allowing for a line voltage of 10% below rated voltage, the voltage drop of the transformer is limited to 5% to ensure continued satisfactory operation. Type E transformers are designed with windings of low

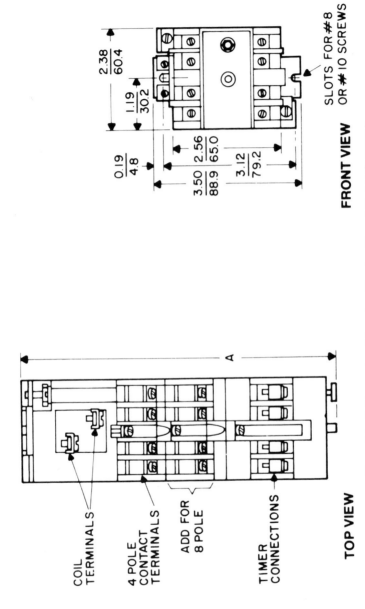

2.38 / 60.4

1.19 / 30.2

0.19 / 4.8

2.56 / 65.0

3.50 / 88.9

3.12 / 79.2

SLOTS FOR #8
OR #10 SCREWS

FRONT VIEW

A

COIL
TERMINALS

4 POLE
CONTACT
TERMINALS

ADD FOR
8 POLE

TIMER
CONNECTIONS

TOP VIEW

Figure 9-1 Solid-state time-delay relay. (Courtesy General Electric Company.)

152

Class 9070
Type EO Transformer

600 V max.
14 sizes — 0.025
through 2.000 kVA
continuous load
capacity

Figure 9-2 Control circuit transformer.

impedance to provide this exact voltage regulation. Good transformer regulation characteristics are essential in order that transformer selection can be made on the basis of continuous VA rating, which is always the most economical choice.

On Square D control transformers, screw-type terminals are accessible from the front and eliminate the terminal blocks or splicing usually required with flexible leads. Each terminal is clearly marked and correct connections for the voltage desired are shown on the nameplate, which is mounted on top of the transformer for easy reference. Transformer coils are layer-wound with insulating paper between layers of wire and are insulated between the primary and secondary windings and also to ground. Maximum temperature rise is limited to 55°C. Windings are of additive polarity.

FUSE BLOCK KITS FOR SECONDARY CIRCUIT PROTECTION

Two bracket-mounted types and two terminal-board-mounted types are available for fusing the secondary of the transformer through 0.750 kVA. One type uses cartridge fuses; the other type uses glass fuses. The advantage of the terminal-board-mounted types is ease of installation, since only the transformer must be mounted on the panel. The bracket-mounted types, however, allow the fusing of both legs of the secondary instead of only one leg.

LIMIT SWITCHES

Limit switches are used to convert a mechanical motion into an electrical control signal. The mechanical motion is usually in the form of a cam, a machine component, or an object moving toward a predetermined position. The cam engages the limit switch lever or plunger, and this in turn makes or breaks an electrical contact inside the switch. This electrical control signal is then used to limit, position, or reverse the machine travel or to initiate another operating sequence. It can also be used for counting, sorting, or as a safety device.

Typical limit-switch applications are in the control circuits of solenoids, control relays, and motor starters which control the motion of machine tools, con-

Figure 9-3 Limit switch used to control electric motors. (Courtesy General Electric Company.)

veyors, hoists, elevators, and practically every type of motor-driven machine. Some different types of limit switches are shown in Fig. 9-3.

Magnet-operated limit switches give accurate position detection of objects at distances up to $3\frac{5}{8}$ in. without physical contact. The switch is actuated by passage of a permanent magnet near the face of the switch. This action causes the normally open contact to close and remain closed until the magnet is removed.

This type of limit switch is especially suited for controlling objects having erratic motion, such as skip hoists, elevators, and overhead conveyors. Most magnet-operated limit switches are designed for a contact rating of 1 A maximum. Some are also available in a solid-state form with a 6-A carry contact rating.

Proximity limit switches are used to detect the presence of metallic objects without physical contact in instantaneous sensing or time-delay applications. They are suited for detecting objects too small to trip conventional limit switches and irregularly shaped objects, and for use in a high temperature. Ferrous or nonferrous metals may be sensed by standard-sized sensing heads at distances up to $\frac{1}{2}$ in., by small remote heads up to $\frac{1}{4}$ in., and by increased range detector up to $2\frac{1}{2}$ in. The ring head sensor will detect ferrous bits as small as 0.004 in.[3], and nonferrous metal bits as small as 0.008 in.[3].

SOLENOIDS

A solenoid (Fig. 9-4) is an electric magnet that applies a straight-line force, in a push or pull motion, when engaged. Typical applications include use on brakes,

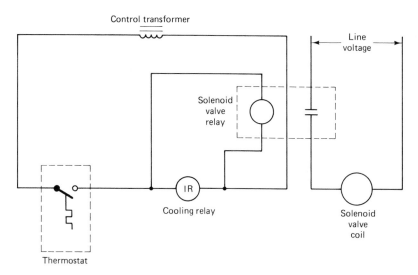

Figure 9-4   A solenoid is an electric magnet that applies a straight-line push or pull when engaged.

conveyors, gates, safety devices, punch presses, clutches, machine tools, door openers, and valves.

In selecting this type of control, the following should be given consideration:

1. Type of operation (push or pull, with or against gravity or horizontal action required)
2. Pounds of force required (match as closely as possible; underrating or overrating will result in short life)
3. Stroke required (in inches)
4. Coil voltage required
5. Type of mounting required

FOOT SWITCHES

A foot switch is a control device operated by a foot pedal used where the process or machine requires that the operator have both hands free. Foot switches usually have momentary contacts but are available with latches that enable them to be used as maintained contact devices.

PRESSURE FLOAT SWITCHES

Industrial pressure switches are designed to cover the wide range of requirements encountered in the control of pneumatic or hydraulic machines. These controls are most commonly employed in the control circuits of welding equipment, machine tools, and high-pressure lubricating systems.

Domestic and commercial pressure switches are designed for the control of electrically driven water pumps and air compressors. Most of these switches are diaphragm actuated and contacts open on increased pressure.

General-duty float switches are designed for the automatic control of ac or dc pump motor magnetic starters or for the automatic direct control of light motor loads. Open tank or sump operation can be obtained by reversing the positions of the float and counterweight.

Domestic sump pump controls are designed specifically for sump pumps or cellar drainers of the small domestic type. They are weight operated and usually include two weights, a chain, and a compensating spring.

Float switches for condensate pumps, as the name implies, are used primarily on condensate pumps. These switches are flange mounted and float movement is transmitted through a bellows seal. One type is attached to the tank by means of a $2\frac{1}{2}$-in. screw-in connection. An external pointer indicates the float position within the tank when the unit is mounted.

SOLID-STATE LOGIC

Solid-state logic is a method of electrically controlling machine or process functions by the use of logic elements and accessories that are not subject to wear and erosion because they have no moving parts. Logic elements perform the same functions that relays perform in conventional control.

Solid-state logic functions accept electrical input signals from sensing devices such as limit switches, pushbuttons, and photocell units, and act as the decision-making portion of the control. When a power output is needed, an output amplifier or an output relay is used to convert the low-level output from a logic function to the power level required. Logic elements perform their functions by use of transistors.

This system of control has the advantage of extremely long life, high reliability, high speed of response, and ability to operate in dirty, dusty, greasy, and damp environments.

Solid-state logic is generally applied where adverse environmental conditions exist that would cause premature failure in conventional control; where the machine duty cycle makes the use of conventional relays impractical due to the high number of operations; where downtime cannot be tolerated and extreme reliability is required; and where fast response is required. Typical applications include machine tools, welding processes, foundry machines and systems, food-processing equipment, petrochemical processes, material handling, lumbermills, and automotive work.

When a need for such a system exists, the manufacturer can usually design and assemble it when a complete written description of the machine function is furnished. The manufacturer will also need a control schematic or present or proposed relay panel. Generally, static control should be considered when the operation is critical or where many relays are being presently used or considered.

chapter ten

Multispeed AC Motor Starters

Ac motors, both squirrel-cage and wound-rotor inductive, and occasionally, synchronous motors, may be arranged with windings that provide two or more speeds. Two-speed motors may have two separate windings or a single winding capable of rearrangement or pole changing. Four-speed motors usually have two two-speed windings. In any case, the different speeds require different switching setups.

Wound-rotor and synchronous motors require switching or pole changing in both primary and secondary (or field) windings. Multispeed motor starters may consist of manually operated drum switches or of magnetic contactors. Standards for connections and markings have changed over the years and have varied with different motor manufacturers. Therefore, when servicing multispeed controllers, actual connections and markings should be obtained from the wiring diagram or from the motor connection plate.

Line-voltage-type multispeed starters are designed to control separate winding and reconnectable winding squirrel-cage motors to operate at two, three, or four different constant speeds, depending on their construction. The use of an automatic starter and proper control station permits greater operating efficiency and offers protection to both motor and machine against improper sequencing or too rapid speed change. Protection against motor overload is provided in each speed category.

Separate winding-type motors have a winding for each speed required. This motor construction is slightly more expensive, but the controller is relatively simple and a wide variety of speeds can be selected.

Consequent-pole-type motors have a single winding for two speeds. Extra winding taps are brought out to permit reconnection for a different number of stator poles. Although the motor is less expensive, the controller is more complicated, and the speed range is limited to a 2 : 1 ratio, such as 600 : 1200 or 900 : 1800.

TORQUE CHARACTERISTICS

Multispeed motors are divided into three groups with regard to torque characteristics. Selection of the proper group depends on the characteristics of the connected load.

1. *Constant torque:* The horsepower output varies directly with the speed. This type is applicable to conveyors, power feed machines, mills, dough mixers, and reciprocating pumps.
2. *Variable torque:* The horsepower output varies as the square of the speed. This type is applicable to drives having fan or centrifugal pump characteristics.
3. *Constant horsepower:* The motor will deliver rated horsepower at all full-load speeds with consequent reduction of torque at the higher speeds. These motors are generally used for driving cutting tools, lathes, spindles, and so on.

Figure 10-1 shows a two-speed two-winding three-phase motor connected to a multispeed motor starter, and Fig. 10-2 shows starters for a two-speed one-winding (consequent pole) constant- or variable-torque three-phase motor.

The speed of a squirrel-cage motor depends on the number of poles of the motor's winding. On a 60-Hz supply, a two-pole motor runs at about 3450 rpm. Motor nameplates are usually marked with actual full-load speeds, but frequently, motors are referred to by their "synchronous speeds"—3600, 1800, and 1200 rpm,

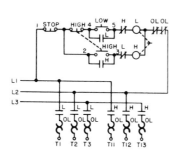

Figure 10-1 Wiring diagram of a two-speed two-winding three-phase motor. (Courtesy Square D Company.)

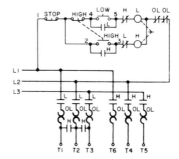

Figure 10-2 Wiring diagram of a two-speed one-winding constant- or variable-torque three-phase motor. (Courtesy Square D Company.)

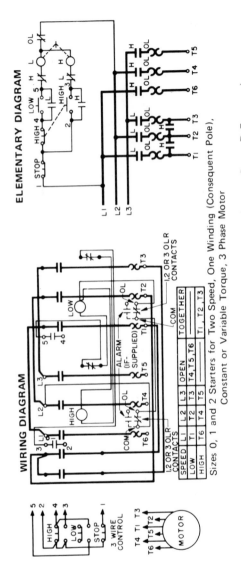

Figure 10-3 Typical ac two-speed motor control circuits. (Courtesy Square D Company.)

Sizes 0, 1 and 2 Starters for Two Speed, One Winding (Consequent Pole),
Constant or Variable Torque, 3 Phase Motor

SPEED	L1	L2	L3	OPEN	TOGETHER
LOW	T1	T2	T3	T4,T5,T6	
HIGH	T6	T4	T5		T1,T2,T3

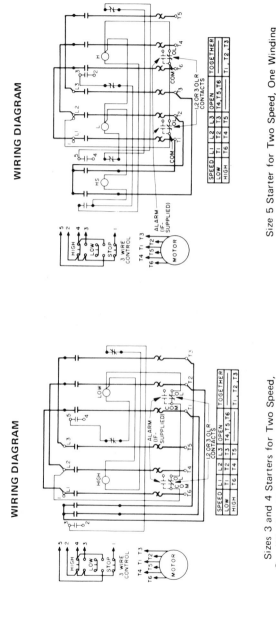

WIRING DIAGRAM

Sizes 3 and 4 Starters for Two Speed,
One Winding (Consequent Pole), Constant
or Variable Torque, 3 Phase Motor

WIRING DIAGRAM

Size 5 Starter for Two Speed, One Winding
(Consequent Pole), Constant or Variable
Torque, 3 Phase Motor

Figure 10-3 (continued)

161

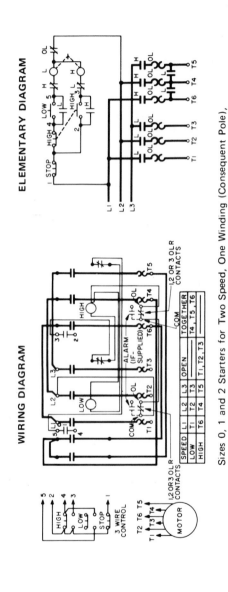

Sizes 0, 1 and 2 Starters for Two Speed, One Winding (Consequent Pole), Constant Horsepower, 3 Phase Motor

Figure 10-3 (continued)

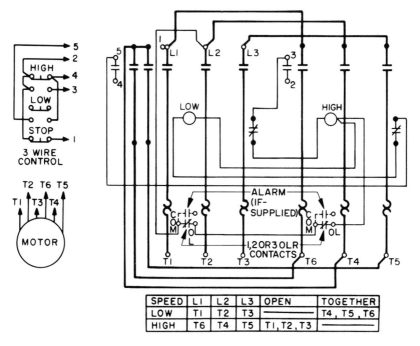

SPEED	LI	L2	L3	OPEN	TOGETHER
LOW	TI	T2	T3	————	T4 , T5 , T6
HIGH	T6	T4	T5	TI, T2, T3	————

Sizes 3 and 4 Starters for Two Speed, One Winding (Consequent Pole),
Constant Horsepower, 3 Phase Motor

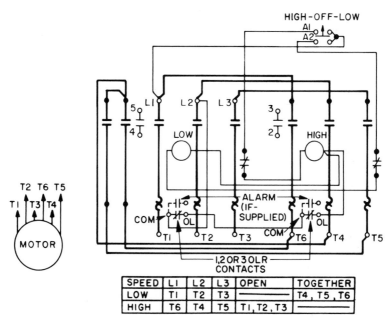

SPEED	LI	L2	L3	OPEN	TOGETHER
LOW	TI	T2	T3	————	T4 , T5 , T6
HIGH	T6	T4	T5	TI, T2, T3	————

Size 0 Starter with HIGH-OFF-LOW Selector Switch (Form C7) for
Two Speed, One Winding (Consequent Pole), Constant Horsepower, 3 Phase Motor

Figure 10-4 Two-speed circuit diagrams for magnetic starters. (Courtesy Square D Company.)

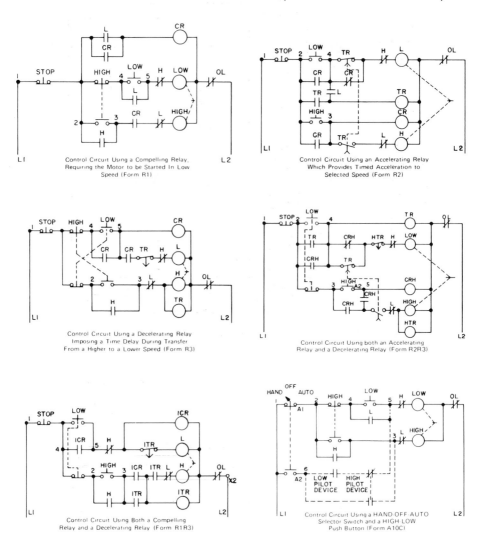

Figure 10-5 Elementary diagrams of special control circuits used with two-speed magnetic starters. (Courtesy Square D Company.)

respectively. Separate windings and/or poles is one way to change the speed of induction motors.

The typical magnetic two-speed controller is used for full-voltage starting of two-speed squirrel-cage induction motors up to 200 hp, 600 V maximum. Overload relays provide overload protection at each speed. Proper motor-speed connections are obtained by adding pushbutton accessories to correspond to each motor speed.

Most two-speed starters are available in NEMA sizes 0 to 5, in open form or

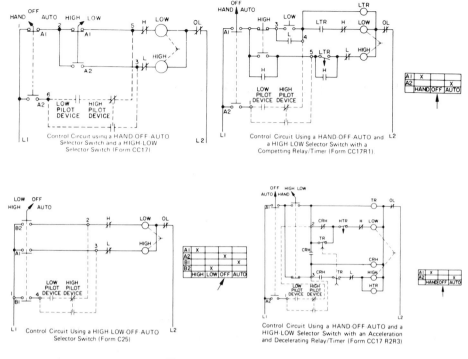

Figure 10-5   (continued)

with type 1 or 12 enclosures and in combination disconnect and circuit breaker forms.

In selecting multispeed motor starters, the following should be given consideration:

1. Type of motor to be controlled (consequent-pole single-winding versus two-winding and separate-winding motors)
2. Fuse-clip size or circuit breaker size on combination form
3. Coil voltage and/or control power transformer application
4. Enclosure type
5. Modifications
6. Overload relay heater selection for each speed (both speeds will not take the same heater element)

Helpful wiring diagrams of multispeed motor controllers are shown in Figs. 10-3 through 10-5.

chapter eleven

Synchronous Motor Controls

Synchronous motors are used in electrical systems where there is need for improvement in power factor or where a low power factor is not desirable. This type of motor is especially adapted to heavy loads that operate for long periods of time without stopping, such as for air compressors, pumps, ship propulsion, and the like.

The construction of the synchronous motor is well adapted for high voltages, as it permits good insulation. Synchronous motors are frequently used on 2300 V or more. Their efficient slow-running speed is another advantage.

A synchronous polyphase motor has a stator constructed in the same way as the stator coils of the induction motor. These are, in turn, grouped to form a three-phase connection, and the three free leads are connected to a three-phase source. Frames are equipped with air ducts that aid the cooling of the windings, and the coil guards protect the winding from damage.

The rotor of a synchronous motor carries poles that project toward the armature, called *salient poles*. The coils are wound on laminated pole bodies and connected to slip rings on the shaft. A squirrel-cage winding for starting the motor is embedded in the pole faces.

The pole coils are energized by direct current, which is usually supplied by a small dc generator called the *exciter*. The exciter may be mounted directly on the shaft to generate dc voltage, which is applied through brushes to slip rings. On low-speed synchronous motors, the exciter is normally belted or of a separate high-speed motor-driven type.

The dimensions and construction of synchronous motors vary greatly depending on the rating of the motors. However, synchronous motors for industrial power applications are rarely built for less than 25 hp or so. In fact, most are 100 hp or more. All are polyphase motors when built in this size. Vertical and horizontal shafts with various bearing arrangements and various enclosures cause wide variations in the appearance of the synchronous motor.

One distinguishing feature of the synchronous motor is that is runs without slip at the synchronous speed determined by the frequency and the number of poles it has.

STARTING SYNCHRONOUS MOTORS

Controls used on synchronous motors have two basic functions: to start the motor and to bring it up to synchronous speed by exciting the dc field. Starting a synchronous motor may be accomplished by any of the across-the-line starters, auto-

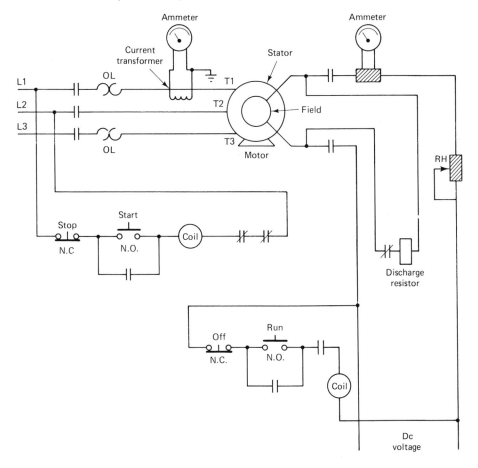

Figure 11-1 Conventional synchronous motor control circuit. In general, starting is a process of accelerating the motor to as high a speed as can be attained on its winding and then applying dc field excitation to obtain synchronization.

transformers, resistance-type motor starters, or other method ordinarily used for induction motors, provided that their capacity is adequate for the motor being started. However, control of the dc field must also be provided to bring the motor up to synchronous speed.

Figure 11-1 shows an elementary diagram of a synchronous motor starting circuit. To operate, the motor is started by depressing the start button, which starts the motor turning. It will start slowly and gradually pick up speed. When the motor has reached its maximum speed, usually indicated by a dial or digital indicator on the instrument panel, the run button is depressed, energizing the coil and thereby closing the dc excitation contact, which opens the field discharge normally closed contacts, and finally, energizes the field.

As mentioned previously, synchronous motors may be used to improve the power factor of an electrical circuit. This is accomplished by creating greater excitation of the field, which creates a leading power factor, which in turn helps a lagging distribution system. However, to utilize this advantage, instruments such as an ammeter, rheostat, or an rpm meter are compulsory. During the speed control, the unity power factor of a synchronous motor may be found by obtaining a minimum reading on the ac ammeter by manipulation of the rheostat. However, such adjustments may cause field currents, or line currents, in excess of the motor rating, and care must be exercised in this manipulation to avoid overload tripping. In other words, the instrument readings should never exceed rated values as shown on the motor nameplate or provided in the manufacturer's specifications.

MOTOR BRAKING

Dynamic braking is used on synchronous motors for quick stopping when a safety bar is operated in the case of emergency. The dc field is left energized, and the motor is disconnected from the ac line and connected to a three-phase resistor. By using quick-acting control sequences and the proper resistors, any size motor can be stopped in approximately 1 second without undue shock or stresses. With dynamic braking, there is no tendency to reverse the motor, and where safety to operators is involved, this feature is often of major importance.

SYNCHRONOUS CONVERTERS

Conversion from ac to dc can be accomplished with a synchronous converter. Usually, the voltage must be decreased with a transformer on the ac side of the synchronous converter because there is a certain fixed ratio between the ac voltage impressed on a synchronous converter and the dc voltage delivered by it. With a single-phase converter, the ac is approximately 70% of the dc voltage. With a three-phase machine, the ac is approximately 60% of the dc voltage. Therefore, to change the dc voltage delivered by the converter, the ac voltage must be varied accordingly.

By changing the field excitation, a converter may be made to correct or compensate for low power factor, as discussed previously. The direct voltage impressed on the line by a synchronous converter may be varied by using a booster such as a small generator or by varying the ac impressed voltage with a potential regulator or a transformer with taps.

SEMIAUTOMATIC SYNCHRONIZING

As discussed in Chapter 9, timing relays may be used in conjunction with synchronous motor controls to bring the motor up to synchronous speed by exciting the field with a definite time delay. This method is shown in Fig. 11-2.

The timing relay in Fig. 11-2 is energized with the main starter coil when the start button is depressed. At the instant the start button is pressed, the normally open interlock closes, but the time relay does not close until the time period for which it is

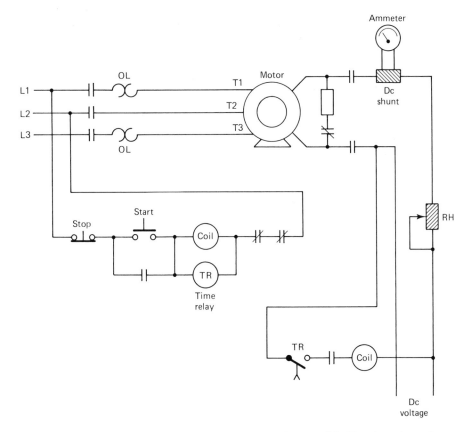

Figure 11-2 Semiautomatic starting of synchronous motors is accomplished by using a timer relay in conjunction with the normal starting controls.

set elapses. During this time, the motor is accelerating to its maximum, and when the predetermined time period has elapsed (calculated to bring the motor speed up to its maximum), the time relay contact closes to accelerate the rotor into synchronism. In using such a system, however, careful adjustments are required initially—observing the instruments at all times—to ensure that the motor will reach full maximum speed in the time allotted; otherwise, the synchronism operation may fail. When this occurs, the stop button must be depressed, and the starting cycle is started all over again.

AUTOMATIC FIELD EXCITATION OF SYNCHRONOUS MOTORS

Automatic application of synchronous motor field excitation is obtained through the use of a polarized field frequency relay. With this method, either across-the-line full-voltage starting or reduced-voltage starting may be used, depending on the characteristics of the motor application. Of the two, across-the-line starting is the more popular method; a magnetic motor starter is used to connect the motor stator directly to the full voltage of the distribution system or at the voltage at which the motor is rated.

Therefore, to start a synchronous motor, voltage is applied to the stator, which causes it to rotate, slowly at first, but with constant acceleration. The rotor is allowed to reach as high a speed as can be attained and then dc voltage is applied to the field for excitation. The part of the synchronous motor control equipment responsible for correctly and dependably applying and removing field excitation is the polarized field frequency relay and reactor.

chapter twelve

Starting and Speed-Regulating
Rheostats and Controllers

Rheostats of the face plate type with self-contained resistors are used in conjunction with manual or magnetic primary control for starting and for speed regulation by secondary control of ac wound-rotor motors ranging from $\frac{1}{4}$ to 25 hp. For reversing service and heavier-duty applications, drum or drum contactor controllers are used with separately mounted resistors. The standard drum controller is used for ratings from $\frac{1}{2}$ to 100 hp; and for the heaviest-duty applications, the drum contactor type is used for ratings from 2 to 300 hp.

MAINTENANCE REQUIREMENTS

Special attention should be given to the maintenance of the secondary control of wound-rotor motors, particularly those used in speed-regulating service, since in a large percentage of applications it is possible for faults to develop through normal wear without causing either immediate shutdown or failure to start. A similar fault in the primary circuit would force immediate correction of the trouble.

Since the motor will continue to start and operate even though an actual open circuit or serious unbalance of resistance may exist in the secondary circuit at certain points on the controller, it is not always understood or appreciated that this condition may result in (1) roasting out of the rotor windings, (2) burning of brushes and collector rings, and (3) overheating of resistors.

Also, undue stress on the equipment may be produced when smooth steps of acceleration provided by control are lost by poor contact or no contact at certain points on the controller. Such conditions may develop without the operator's noting or reporting any difficulty until serious breakdown occurs, at which time it will be recalled that the operator "did have to notch the controller up a step" or that "it jumped a bit on that point."

A definite and regular inspection schedule is essential not only for these reasons, but also because this class of apparatus is of such rugged design and construction that it requires a minimum of attention and may therefore be neglected.

GENERAL MAINTENANCE

To ensure against service interruptions and to keep all types of secondary controllers in good operating condition requires only regular inspection and cleaning and maintenance of the contacts. Some arcing and burning of contact-making parts is unavoidable, but these should be kept smooth to ensure positive low-resistance contact at all times. Occasional dressing with a file may be necessary. The contacts should be lightly lubricated with Vaseline after dressing and cleaning.

Resistors can be readily checked for continuity by testing across rheostat contacts or controller fingers or preferably by raising brushes at the motor collector rings and connecting a test lamp, ohmmeter, or other test equipment across the brush holders or outgoing leads and moving the secondary controller through its full sequence step by step. If repeated across each phase, this will verify continuity of resistors and tap connections as well as indicate any open contacts in the controller. However, since the values of secondary resistance are relatively low, actual measurements of resistance values on each step are not possible with the equipment available in many plants. Here more careful visual inspection must be made to locate and correct loose connections or low-pressure contacts. Tap connections on grid-type resistors, if found movable by hand or showing evidence of heating, should be removed, cleaned, replaced, and pressure of contact increased by taking up on the pressure nuts at the ends of the grid assembly. On ribbon-wound resistors, the clamp connections should be tightened if not solid.

NEW INSTALLATIONS

On new installations, at least two inspections should be made at short intervals after placing in service to ensure that all contacts remain tight after heating and cooling of resistors in service. Since a large amount of heat energy is liberated from resistors, equipment must be located to allow adequate air space and ventilation. Underwriters' Laboratories guidelines and local building code requirements should be followed.

FACEPLATE TYPES

Typical assembly parts lists and wiring diagrams are shown in Figs. 12-1 through 12-4. Rheostats for starting duty are provided with a return spring and latch only in the full-speed position, so the arm cannot be left on intermediate positions. When

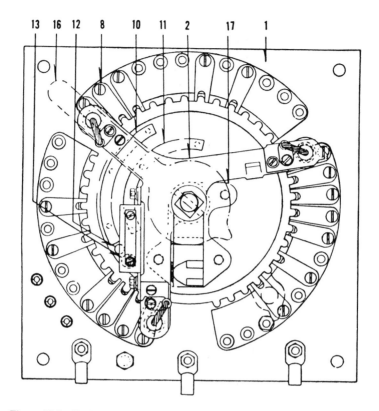

Figure 12-1   Typical faceplate controller and parts list. (Courtesy Westinghouse.)

inspected, these types should be checked to see that the arm returns freely to the start position except when latched, since if operated at an intermediate position, the resistors will be overloaded. Rheostats for speed-regulating duty are designed for continuous operation on any step and the return spring and latch are omitted.

Most faceplate types are designed with reversible stationary contacts which may be turned over when they become worn or burned. When turning them over, all surfaces should be dressed to remove oxide and ensure positive contact. The surface level of adjacent contacts should be checked to ensure that the moving contact will bridge with firm pressure on each stationary contact.

DRUM CONTROLLERS

A diagram of a typical drum controller is shown in Fig. 12-5. The finger support assembly may readily be removed by taking out two bolts to permit convenient dressing of the contact tips.

Two types of drum assembly are used. Lower ratings have contact supporting disks assembled on an insulated steel shaft between the insulating collars, with the

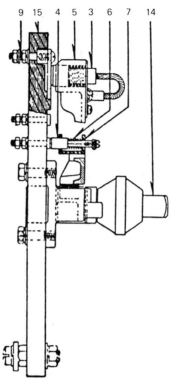

Reference number	Name of part	Number per unit
1	Face plate complete	1
2	Contact arm complete with contacts	1
3	Main contact with shunt	3
4	Auxiliary contact	2
5	Main contact spring	3
6	Auxiliary contact spring	2
7	Insulation channel for auxiliary contact	1
8	Stationary contact	21
9	Stationary contact stud	21
10	Contact segment — outer	1
11	Contact segment — inner	1
12	Stationary contact button	1
13	Stationary contact lava button	1
14	Shaft assembly	1
15	Base	1
16	Handle	1
17	Bearing	1

Figure 12-2 Sectional view of a faceplate controller. (Courtesy Westinghouse.)

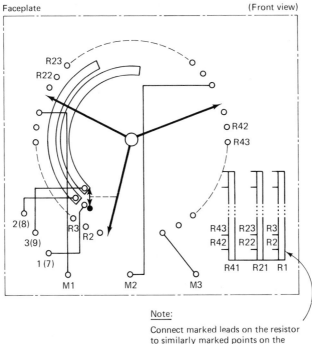

Faceplate   (Front view)

2(8)
3(9)
1 (7)
M1   M2   M3

Note:

Connect marked leads on the resistor
to similarly marked points on the
faceplate except R1-R21-R41, which
are common on the resistor.

Figure 12-3 Typical wiring diagram for a faceplate controller. (Courtesy Westing-house.)

reversible contacts bolted directly to the supporting disks. Larger ratings, 150 to 300 A, have heavy curved copper segment plates supported by molded insulating supports bolted to a steel shaft. On these types, the upper and lower segments have the same shape, to permit interchange and reversal. All drum segments on all types are reversible so that the old trailing edge becomes the leading edge when turned 180°. This not only provides double life but also materially simplifies maintenance.

The fingers should be adjusted so that they will drip not more than $\frac{1}{8}$ in. below the surface of the drum contacts. Bearings, star wheel, and pawl should be regularly cleaned and oiled.

DRUM CONTACTOR CONTROLLERS

Drum controllers are used for starting and speed regulation on certain types of ac motors. The drum contactor controller, for example, is designed for the most severe operating conditions, and consists of a series of contactors that are closed by cams

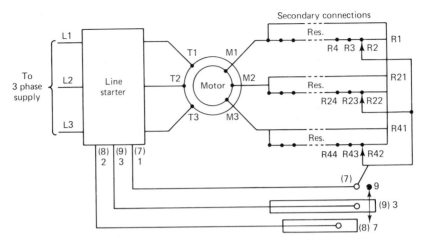

Figure 12-4   Elementary controller diagram. (Courtesy Westinghouse.)

on the operating shaft and open by positive spring pressure. The contacts are the same as those used on corresponding sizes of magnetic contactors and are of the well-known rolling type. This rolling action limits all arcing and burning to the contact tips, where current is carried only momentarily, leaving the actual current-carrying contact surfaces in perfect condition. Contacts need replacement only when they are burned so that little or no rolling action is left. When installing new contacts, they should be adjusted so that the proper rolling action is obtained.

RESISTORS FOR AC MOTORS

All standard ac wound-rotor motors, whether for two- or three-phase circuits, have their secondaries wound for three-phase operation. The resistors for each phase used with these motors are identical with the exception of the terminal marking. The resistor for the first phase has its terminals marked consecutively R-1, R-2, R-3, and so on; the second phase, R-11, R-12, R-13, and so on; and the third phase, R-21, R-22, R-23, and so on. The actual resistor will consist of one, two, three, or multiples of three frames of tubes or grids. Check the nameplate to see if all frames have been received. When two frames are furnished, they should be connected in series by connecting terminals A to A. When three frames are furnished, this connection is not required. When more than three frames are supplied, sort out the frames for each phase according to the terminal marking, and connect those frames belonging to each phase by connecting A to A, B to B, and so on. Make all connections in line with the following information and the diagram located on the controller cover.

Secondary resistors for ac motors are designed for star connection. Resistors for most manual controllers may be connected wth all three secondary phases closed

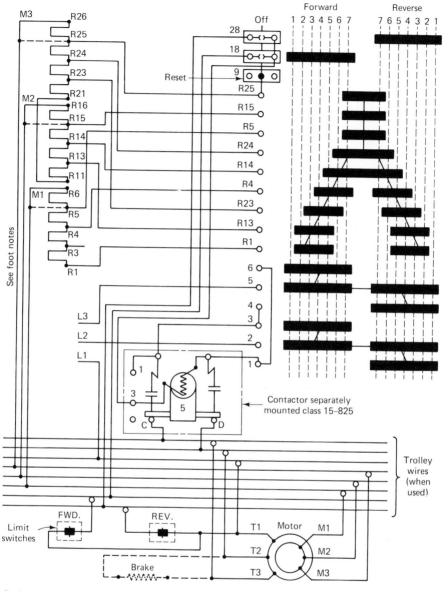

Resistor connections: Read the instruction card which is attached to the resistors. The connections shown in these diagrams are for a starting motor with one phase of its secondary open on the first point of the controller. When higher starting torque is desired, or when connecting motors are rated above 80 hp, connect R1 to R11 at the resistor and finger marked R1 on controller to terminal R3 on resistors. Resistor steps R5 to R6, R15 to R16, and R25 to R26 are for resistance which remains permanently in the circuit. When these are not supplied, connect M1, M2, and M3 to R5, R15 and R25 as indicated by the dashed lines.

Figure 12-5 Typical wiring diagram for a drum controller. (Courtesy Westinghouse.)

TABLE 12-1 AES Service Classification Table

Percent of Full-Load Current on First Point	Starting Torque (% of Full Load)[a]					Resistor Class Number				
				Wound-Rotor Induction Motors		Starting Duty		Intermittent Regulating Duty[b]		Continuous Regulating Duty[c]
	Series Motors	Compound Motors	Shunt Motors	One-Phase Starting	Three-Phase Starting	Light—15 Sec on Out of 4 Min	Heavy—30 Sec on Out of 4 Min	Light—1 Min on Out of 4 Min	Heavy—2 Min on Out of 4 Min	
25	8	12	25	15	25	11	31	51	71	91
50	30	40	50	30	50	12	32	52	72	92
70	50	60	70	40	70	13	33	53	73	93
100	100	100	100	55	100	14	34	54	74	94
150	170	160	150	85	150	15	35	55	75	95
200	250	230	200	—	200	16	36	56	76	96

[a]Based on Westinghouse motors.

[b]The letter D added to classes 52 or 72 indicates additional capacity for dynamic lowering.

[c]The letter V added to class 92 indicates resistor designed for varying torque applications where the horsepower varies as the cube of the speed.

Source: Courtesy Westinghouse.

or with one secondary phase open on the first point of the controller. Resistors for magnetic controllers are connected with all three phases closed in the secondary on the first point.

The torque obtained with a resistor of a given class number varies with the connection used on the first point of the controller. The torques available on the first point with single-phase and three-phase starting are listed in Table 12-1. Where it is possible to use both methods of connection, the control diagram shows one method of connection and explains how to obtain the other method. The method actually shown on the diagram is ordinarily recommended, but if a change in starting torque is desirable, the other method may be used.

STARTING AND SPEED-REGULATING RHEOSTATS FOR DC MOTORS

Rheostats are used for starting and speed regulation of series, shunt, and compound wound motors in nonreversing service for the operation of fans, blowers, pumps, machine tools, and similar dc motor applications, ranging from $\frac{1}{4}$ to 150 hp.

CLEANING

To keep rheostats in good operating condition, periodic inspection and cleaning and smoothing the contacts with a file are usually all that is needed. The low upkeep cost of rheostats is due to such features as magnetic blowout devices, high contact pressures, contacts raised above the faceplate, rugged moving parts, and easy accessibility. However, some arcing and burning of contact-making parts is unavoidable, and dressing with a file may be required occasionally. Contacts should always be smooth. After each treatment with a file, all parts should be thoroughly cleaned, including surfaces between contacts, and contacts should be very lightly greased with Vaseline. Sometimes, cutting of the metal is caused by sharp contact edges or by abrasive matter in the air. If the latter, greasing should be omitted.

REVERSING CONTACTS

Most types of rheostats have movable and stationary contacts that can be turned over and used on the other side. This gives the contacts double life, but turning over should be done only when abnormal burning and subsequent dressing with a file made adjacent contact surfaces uneven. The moving contact must bridge and be in firm contact with each adjacent stationary contact. If these points are not checked, irregular increments in speeds or voltage will be the result. On larger rheostats or those with movable contacts of the compensated type, a slight variation between surfaces will not impair operation. It is generally advisable when turning over one contact to turn all others over also.

SPRING TENSION

Firm pressure between moving and stationary contacts should be maintained by proper spring tension to minimize pitting, heating, and oxidation, which aggravate abnormal conditions. The recommended moving contact spring pressure for each type of rheostat may be checked with a hook spring balance. The pull is measured in pounds needed to separate the contacts. In most cases this is impractical. The desired limit on spring adjustments is to provide pressure not great enough to injure the surface of either moving or stationary contacts or to set up a frictional force that prevents resetting the arm on rheostats of the low-voltage-release type. The turns of the spring should not touch when compressed. Weak springs should be discarded.

RESISTOR REPLACEMENT

Abnormal starting or operating conditions may burn out a section of the rheostat resistor. In this case, the faceplate—together with the resistors—can easily be removed from the box and the damaged units repaired or replaced. If the resistor is of the wire-wound bobbin type, covered with a cement coating, a complete new set is recommended; if it is a suspension grid type, the burned-out section should be replaced with one of the same pattern or style number; if it is a type M edge-wound strap resistor, the break may be bridged with a clamp similar to the type furnished with the resistor. In most types of rheostats, the field resistors are removable from the resistor mounting without disconnecting the wiring at the faceplate.

MAGNET COIL

Practically all starting rheostats of low horsepower rating have a magnet coil as part of the low-voltage release. This coil is connected directly across the line. The release is adjusted to hold the operating handle in the last running position as long as the voltage is normal.

If it is desired to hold the starter in at slightly less than normal voltage, the holding power of the coil can be increased by filing down the little brass pin on the moving arm. However, this method should not be carried too far, nor should the operating arm be held in the running position by force. It is more economical to install a new magnet coil or spring than to take a chance on harming the motor by excessive starting current.

The rheostat should not be used to stop the motor. A safety switch or circuit breaker is provided for this purpose.

CHECKING LOOSE CONNECTIONS

After a rheostat has been installed and is in operation, all connections should be checked at least twice to make sure that they remain tight when subject to heating. Loose connections are a considerable source of trouble, causing many delays, and they are difficult to find unless burning or failure of some other device occurs. Means of checking are limited and in most cases restricted to mechanical inspection. This should be done periodically. One of the best devices for checking rheostats is an ohmmeter. It is convenient, rugged in construction, and small in size. Ohmic values are indicated on the dial. With this instrument and a record of normal resistance value, any change between two rheostat points can readily be determined. From the readings it will be obvious if the circuit is normal or not. Incidentally, burned-out resistor tubes can be traced quickly in this manner.

INSTALLATION

Rheostats should always be mounted so that the ventilating hood is at the top. An air space should be allowed both above and below for ventilation. Always check Underwriters' Laboratories regulations and local building codes for any special requirements. All wiring should be done in accordance with the NEC.

chapter thirteen

Direct-Current Controllers

Direct-current controllers are classified as many different types, but essentially they are either manually or automatically operated. Small dc motors of, say, less than $\frac{1}{2}$ hp consume very little current upon starting and therefore can be started by placing full voltage across the motor terminals. Large dc motors cause large initial currents to flow because they have a low resistance, and the excessive current flow during starting may damage the motor or trip the overcurrent device.

To start a large dc motor, it is necessary to place a resistance unit in series with the motor so that the starting current is reduced to a safe value. As the motor accelerates, this resistance can gradually be decreased. Once the motor reaches the desired speed, the resistance is no longer necessary because the motor is now generating a voltage that is in opposition to the impressed voltage, thereby preventing excessive current flow. This opposing voltage is called the *counterelectromotive force* (CEMF), and its value will depend on the speed of the motor, which is greatest at full speed and zero at standstill.

In general, dc motors of less than 2-hp rating can be started across the line, but with larger motors it is usually necessary to put resistance in series with the armature when it is connected to the line. This resistance, which reduces the initial starting current to a point where the motor can commutate successfully, is shorted out in steps as the motor comes up to speed and the countervoltage generated is sufficient to limit the current peaks to a suitable value. Accelerating contactors that sort out successive steps of starting resistance may be controlled by countervoltage or by definite-time relays. For small motors used on auxiliary devices the CEMF starter is satisfactory. The definite-time starter is more widely used, however, and has the advantage of being independent of load conditions.

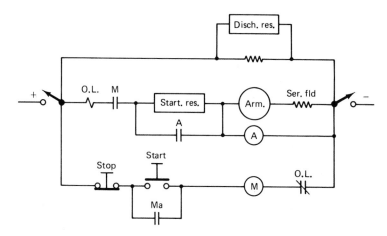

Figure 13-1 Basic requirements of a nonreversing dc motor starter.

The following diagrams illustrate some of the circuits commonly used for dc motor control. Basic requirements of a nonreversing dc starter in its simplest form is shown in Fig. 13-1. When the start pushbutton is depressed, line contactor M closes, energizing the motor armature through the starting resistance. As the motor comes up to speed, the countervoltage and the voltage across the motor armature and series field increase. At a predetermined value the accelerating contactor A closes, shorting out the starting resistance.

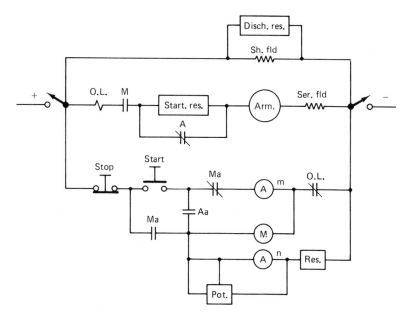

Figure 13-2 Diagram for nonreversing constant-speed definite-time dc starter.

Figure 13-2 shows a typical nonreversing constant-speed definite-time starter. The accelerating contactor is equipped with a time-delay mechanism. This contactor A is of the magnetic-flux-decay type. It is spring-closed, equipped with two coils, and has a magnetic circuit that retains enough magnetism to hold the contactor armature closed and the contact open indefinitely. Main coil Am has sufficient pull to pick up the armature and produce permanent magnetization. Neutralizing coil An is connected for polarity opposite the main coil. It is not strong enough to affect the pickup or holding ability of the main coil, but when the latter is deenergized, the neutralizing coil will buck the residual magnetism so that the contactor armature is released by the spring, and the contacts close. By adjusting the potentiometer, the voltage impressed on this coil, and hence the time required for the contactor to drop out, can be varied. When the start button is depressed, accelerating contactor coil Am is energized, causing contact A to open and auxiliary contact Aa to close. Contact Aa energizes line contactor M, and normally open auxiliary contacts Ma establish a holding circuit. Neutralizing coil An is also energized. The opening of contact Ma deenergizes coil Am, and contactor A starts timing. At the set time the main normally closed contacts on A close, shorting out the starting resistance and putting the motor across the line.

The same kind of starter as in Fig. 13-2 but designed for use with a motor of larger horsepower is shown in Fig. 13-3. This starter provides two steps of definite-

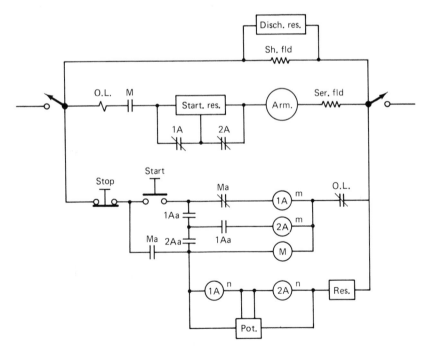

Figure 13-3 Same type of starter as that shown in Fig. 13-2 but designed for motors with larger horsepower.

time starting. The operation is essentially the same as in Fig. 13-2, but the first accelerating contactor 1A does not short out all the starting resistance. It also starts 2A timing, which finally shorts out the remaining resistance. The normally open auxiliary contacts on the accelerating contactors in Figs. 13-2 and 13-3 are arranged so that it is necessary for the accelerators to pick up before the line contactor can be energized. This is a safety interlocking scheme that prevents starting the motor across the line if the accelerating contactors are not functioning properly.

One way of producing dynamic braking is shown in Fig. 13-4. Control circuits have been omitted, since they are a duplicate of those shown in Figs. 13-2 and 13-3. Line contactor M has two poles, one normally open and the other normally closed. Both poles are equipped with an operating coil and are on the same armature, which is hinged between the contacts. In starting, when line contactor M closes, normally closed contact Ma opens. When the stop button is depressed, the line contactor drops out and contact Ma closes. The motor, now acting as a generator, is connected to the braking resistor, and coil Ma is energized by the resultant voltage. It causes M to seal in tightly, establishing good contact pressure and preventing the contact from bouncing open.

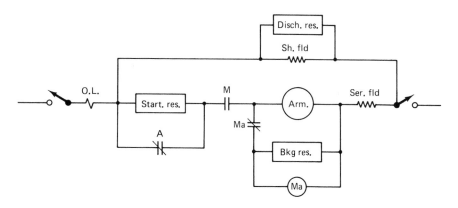

Figure 13-4   One method of producing dynamic braking on a dc motor.

In the more modern controllers a separate spring-closed contactor is used for dynamic braking (see Fig. 13-5). Operation is similar to that described for Fig. 13-2, except that the energizing of coil Am and the picking up of accelerating contactor A and closing contact Aa energize dynamic braking contactor DB, which in turn energizes line contactor M through its auxiliary contact DBa. This arrangement not only ensures that the dynamic braking contactor is open but also that it is open before the line contactor can close. To obtain accurate inching, such as is required for most machine tool drives, the motor must respond instantly to the operation of the pushbutton. In the scheme shown in Fig. 13-5, the closing of the line contactor is delayed until the accelerating contactor and the dynamic braking contactor pick up.

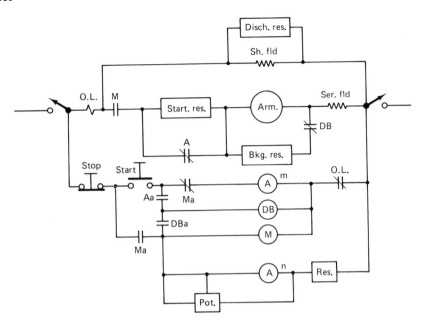

Figure 13-5   Separate spring-closed contactors are used in some applications of dynamic braking.

Figure 13-6 shows an arrangement to secure quicker response of the motor for more accurate inching. Accelerating contactors 1A and 2A are energized in the "off" position. Hence, when the start button is depressed, the dynamic braking contactor picks up immediately, and its auxiliary contact DBa picks up M line contactor.

One method of connecting the full-field relay, used with adjustable-speed motors having a speed range in excess of 2:1, is shown in Fig. 13-7. Coil FF is energized by the closing of the normally open auxiliary contact Aa and remains closed until the last accelerating contactor drops out. The contactors of the full-field relay FF are connected to short out the field rheostat, thereby applying maximum field strength to the motor during the starting period.

Another method of applying the full-field relay is shown in Fig. 13-8. This arrangement ensures a full field on starting and provides for limiting the armature current when the motor is accelerating from the full-field speed to the speed set by the rheostat. Field accelerating relay FA is equipped with two coils, one a voltage coil connected across the starting resistance and the other a current coil connected in series with the motor armature (see Fig. 13-2 for the remainder of the circuit). When line contactor M closes, the voltage drop across the starting resistor is practically line voltage, and relay FA is picked up quickly. When accelerating contactor A closes, voltage coil FAv is shorted, but the closing of A produces a second current peak, and current coil FAc holds relay FA closed. As the motor approaches full-field speed, this current decays and allows the FA contacts to open, weakening the

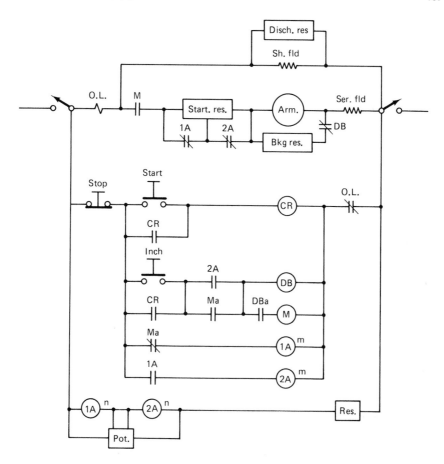

Figure 13-6 Control for inching a dc motor.

motor field. When the motor attempts to accelerate, the line current again increases. If it exceeds the pickup value of coil FAc, the relay will close its contacts, arresting acceleration and causing a decay of line current, which again causes FA to drop out. High inductance of the motor field plus inertia of the motor and drive prevent rapid changes in speed. Hence the motor will not reduce its speed, but the increased field current will reduce the armature current and cause FA to drop out. The fluttering action will continue until the motor reaches the speed set by the rheostat. Setting the FA relay current coil determines the maximum current draw during this part of the acceleration period. Since relay FA must handle the highly inductive field circuit, a good blowout arrangement is necessary. Hence the relay is usually equipped with a shunt blowout coil, FABO.

Connection of a field loss relay, to prevent excessive speed if the shunt field is deenergized while voltage remains on the armature, is shown in Fig. 13-9. A field loss relay usually consists of a current relay in series with the motor shunt field and

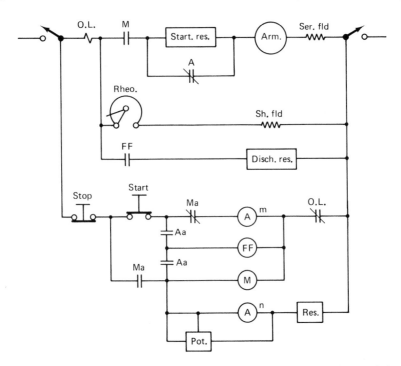

Figure 13-7 Method of connecting full-field relay used with adjustable-speed dc motors.

is adjusted to pick up on full-field current and remain closed at any current within the operating range of the motor field current. The contacts of relay FL are connected in series with the overload relay contacts so that the opening of its contacts will deenergize the control by opening the line contactor. This type of field loss protection does not protect against the possibility of a short circuit across a part of the field, say across the one field coil. This would cause the motor speed to rise considerably, but the current in the field circuit would also rise. Consequently, the series current relay would not respond.

VARIABLE-VOLTAGE CONTROL

The variation in speed obtainable by field control on the conventional dc motor will not, in most cases, exceed 4 : 1, due to the sparking difficulties experienced with very weak fields. Although the range may be increased by inserting resistance in series with the armature, this can be done only at the expense of efficiency and speed regulation.

With constant voltage applied to the field, the speed of a dc motor varies directly with the armature voltage; therefore, such a motor may be steplessly varied

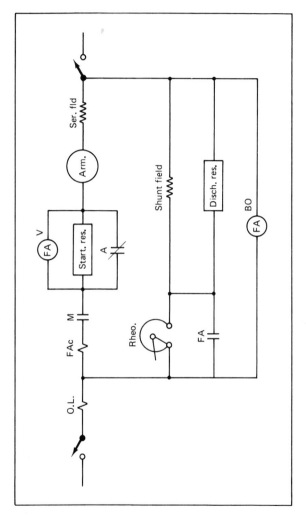

Figure 13-8 This wiring arrangement ensures full field on starting.

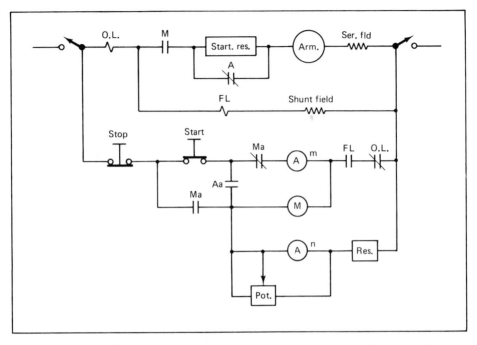

Figure 13-9 Wiring arrangement to prevent excessive motor speed should the shunt field be deenergized while voltage remains on the armature.

from zero to maximum operating speed by increasing the voltage applied to its armature. The diagram in Fig. 13-10 shows the arrangement of machines and the connections used in one type of variable-voltage control designed to change speed and reverse rotation. The constant-speed dc generator (B) is usually driven by an ac motor (A), and its voltage is controlled by means of a rheostat (R). Note that the fields of both generator and driving motor are energized from a separate dc supply or by an auxiliary exciter driven off the generator shaft, causing the strength of the motor field to be held constant, while the generator field may be varied widely by rheostat R.

With the system in operation, generator B is driven at a constant speed by prime mover A. Voltage from B is applied to the dc motor (C), which is connected to the machine to be driven. By proper manipulation of rheostat R and field-reversing switch S, the dc motor may gradually be started, brought up to and held at any speed, or reversed. As all of these changes may be accomplished without breaking lines to the main motor, the control mechanism is small, relatively inexpensive, and less likely to give trouble than is equipment designed for heavier currents.

The advantage of this system is the flexibility of the control and the complete elimination of resistor losses, the relatively great range over which the speed can be varied, the excellent speed regulation on each setting, and the fact that changing the

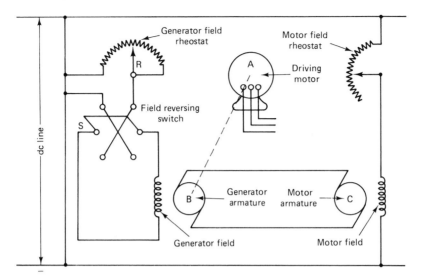

Figure 13-10   Variable-voltage control used in special applications of dc motors.

armature voltage does not diminish the maximum torque the motor is capable of exerting, since the field flux is constant.

By means of the arrangement shown in Fig. 13-10, speed ranges of 20 : 1— compared to 4 : 1 for shunt field control—may be secured. Speeds above the rated normal full-load speed may be obtained by inserting resistance in the motor shunt field. This represents a modification of the variable-voltage control method, which was originally designed for the operation of constant-torque loads up to the rated normal full-load speed.

Because three machines are usually required, this type of speed control finds application only where great variations in speed and unusually smooth control are desired. Steel mill tools, electric shovels, passenger elevators, machine tools, turntables, large ventilating fans, and similar equipment represent the type of machinery to which this method of speed control has been applied.

OVERLOAD RELAYS

To protect the motor and related circuits from accidental or prolonged overloads, either the starter or the motor should be equipped with automatic devices that will open the circuit should an overload occur. This protection can be provided by fuses, circuit breakers, or overload relays.

Overcurrent protection must be provided in the line of every motor circuit, but additional protection should be provided in the form of magnetic overload relays. These are used in both manual and automatic starters.

Figure 13-11 shows a plunger type of overload relay. When the current through the coil reaches the value set by the adjustable screw, the plunger is drawn

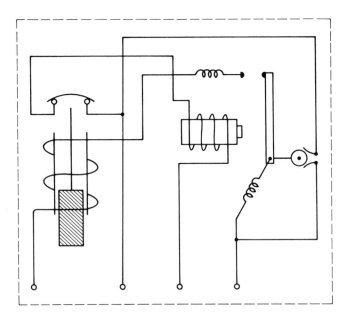

Figure 13-11   Plunger type of overload relay.

up and opens two contacts. This relay can be used on both manual and automatic controllers.

Most overload relays used on modern starters are thermally operated and usually consist of two strips of metal having different degrees of thermal expansion which are welded together. If this bimetallic strip is heated, it will deflect sufficiently to trip two normally closed contacts, which in turn will open-circuit the holding coil of a magnetic contactor, causing the main contacts to open. An advantage of this protective device is that it provides a time delay which prevents the circuit from being opened by momentary high starting currents and short overloads. At the same time, it protects the motor from prolonged overloading.

chapter fourteen

Connection Diagrams for Pushbutton Circuits

Standard-duty control stations are used with magnetic motor starters to govern the starting, stopping, or reversing of all types of electric motors. Pushbutton units are assembled in various combinations to form unified stations for remote control of starter and motor operations. Standard-duty stations may consist of one, two, or three control units and can be furnished in either surface-mounted or flush-mounted construction. Both momentary- and maintained-contact pushbutton assemblies are available in two-unit stations.

JOGGING CIRCUIT

The jogging connection shown in Fig. 14-1 is used primarily when machines must be operated momentarily for inching, such as in a machine tool set up for maintenance. The jog circuit energizes the starter only when the jog button is depressed, thereby giving the machine operator instantaneous control of the motor drive. In the circuit under consideration, when the jog button is depressed, the control relay is bypassed, and the main contactor coil is energized solely through the jog button; when the jog button is released, the contactor coil releases immediately. Pushing the start button closes the control relay and is held in by its own normally open contacts. The main contactor coil is in turn closed by another set of normally open contacts on the control relay and is held in the "on" position.

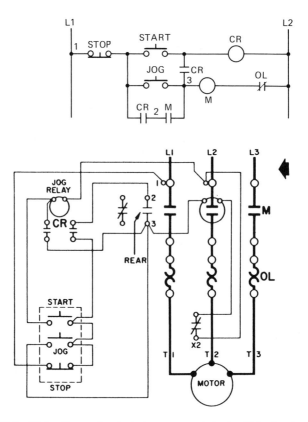

Figure 14-1   Wiring diagram for a typical motor control circuit utilizing jogging feature. (Courtesy Cutler-Hammer Products.)

Utilizing Auxiliary Contacts

The circuit in Fig. 14-2 shows two motors with their controls arranged in such a manner that one cannot be started until the other is running. This type of control is necessary when one machine feeds a second, such as in a conveyor system. To prevent the first machine from piling up material on the second, the second machine is started first. This is accomplished by interconnecting the pushbutton stations. The control circuit of the second starter is wired through the auxiliary contacts of the first; this prevents it from starting until after the first starter is energized. A timer can be connected between the starters so that the second motor will run a short time after the first motor has stopped.

Controlling Motor from Several Locations

The circuit shown in Fig. 14-3 finds application where a single motor is operated from several remote locations. This diagram also suggests other circuit possibilities, including multiple emergency stop stations. A typical application is

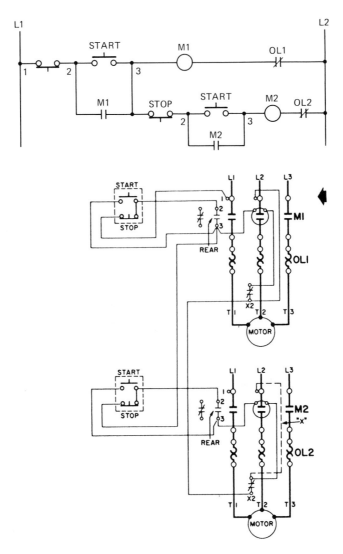

Figure 14-2 Two pushbuttons arranged so that one motor cannot be started until the other is running. (Courtesy Cutler-Hammer Products.)

when the operator of a machine is working in a location that is not within easy reach of the main pushbutton station.

A circuit that is the opposite of that shown in Fig. 14-3 appears in Fig. 14-4, which shows the operation of three motors from a single pushbutton location. In this circuit, the starters are wired so that all three will automatically be disconnected from the line if a maintained overload occurs on any one starter. This is accomplished by wiring the holding circuit of each starter through the auxiliary contacts

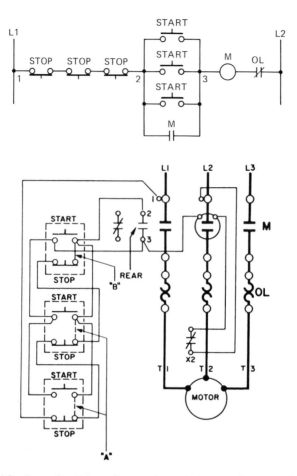

Figure 14-3   Connection diagram for operating a single motor from several different remote locations. (Courtesy Cutler-Hammer Products.)

of one of the other two. Since the control circuit is common among the starters, incoming power lines to all three starters must be opened by a disconnect preceding each of the starters to disconnect the starters completely from the line.

FLOAT SWITCH CIRCUIT

A common single-pole double-throw float switch circuit is shown in Fig. 14-5; this is the type used in sump pump and tank-filling applications. The circuit shown is connected for tank operation, but for sump pump operation, the float switch can be connected as indicated by the dashed lines in the drawing. In both cases, the float

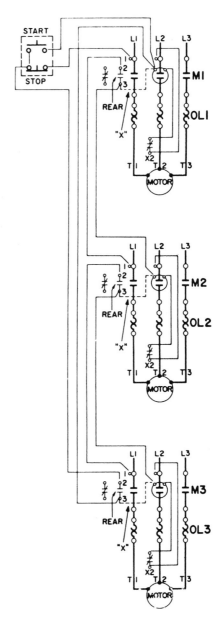

Figure 14-4 Connection diagram for operating three motor starters from a single push-button location. (Courtesy Cutler-Hammer Products.)

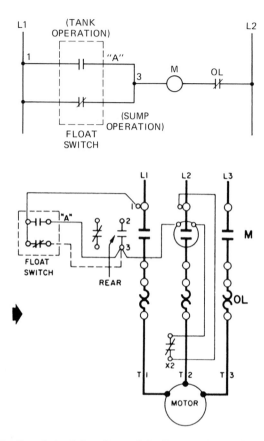

Figure 14-5   Control circuit for a float switch. (Courtesy Cutler-Hammer Products.)

within the sump pump or the tank controls the opening and closing of the contacts in the switch, which in turn controls the motor starter.

SOLID-STATE CONTROL CIRCUITS

The wiring diagrams shown in Figs. 14-6 through 14-10 show various connections of their solid-state devices. A careful study of these should enable anyone with some training in blueprint reading to understand their functions.

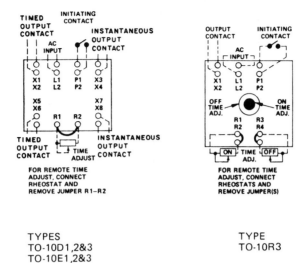

TYPES
TO-10D1,2&3
TO-10E1,2&3

TYPE
TO-10R3

ON DELAY or, INTERVAL, MAINTAINED START   **OFF DELAY**

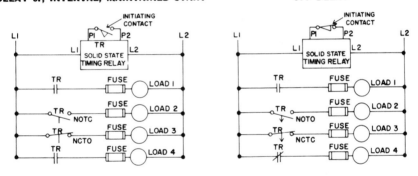

REPEAT CYCLE

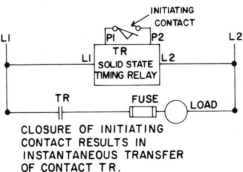

CLOSURE OF INITIATING
CONTACT RESULTS IN
INSTANTANEOUS TRANSFER
OF CONTACT TR.

Figure 14-6 Solid-state timing relays. (Courtesy Square D Company.)

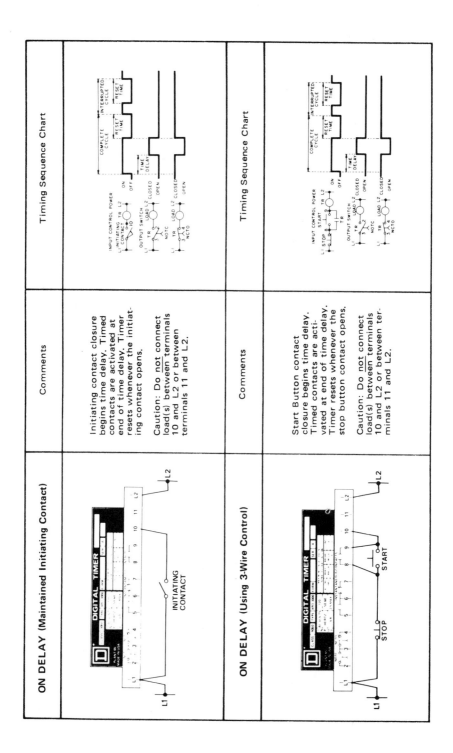

ON DELAY (Maintained Initiating Contact)	Comments	Timing Sequence Chart
	Initiating contact closure begins time delay. Timed contacts are activated at end of time delay. Timer resets whenever the initiating contact opens. Caution: Do not connect load(s) between terminals 10 and L2 or between terminals 11 and L2.	
ON DELAY (Using 3-Wire Control)	Comments	Timing Sequence Chart
	Start Button contact closure begins time delay. Timed contacts are activated at end of time delay. Timer resets whenever the stop button contact opens. Caution: Do not connect load(s) between terminals 10 and L2 or between terminals 11 and L2.	

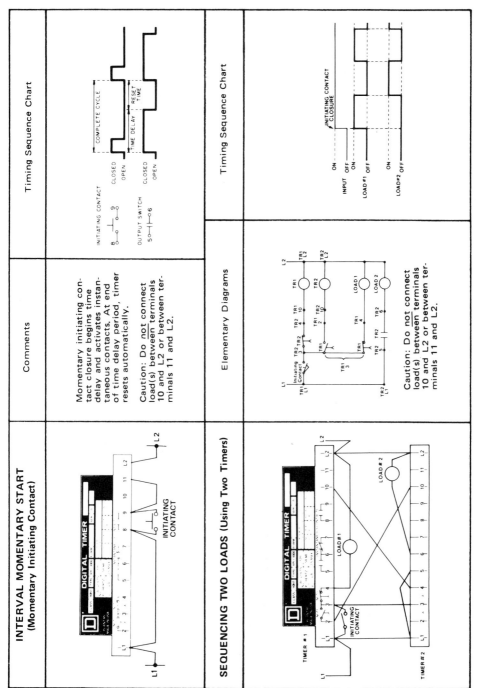

Figure 14-7 Digital solid-state timer. (Courtesy Square D Company.)

OFF DELAY	Comments	Timing Sequence Chart
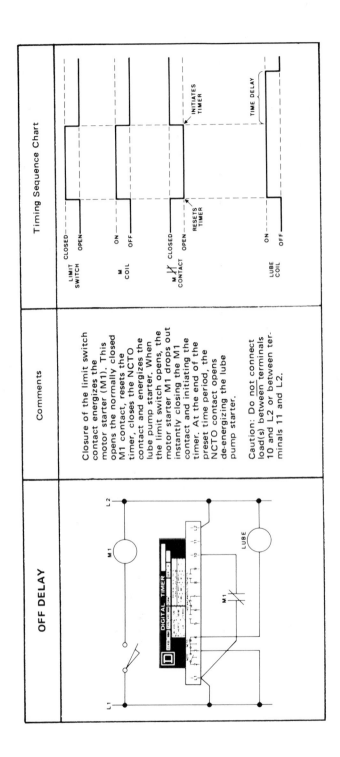	Closure of the limit switch contact energizes the motor starter (M1). This opens the normally closed M1 contact, resets the timer, closes the NCTO contact and energizes the lube pump starter. When the limit switch opens, the motor starter M1 drops out instantly closing the M1 contact and initiating the timer. At the end of the preset time period, the NCTO contact opens de-energizing the lube pump starter.

Caution: Do not connect load(s) between terminals 10 and L2 or between terminals 11 and L2. | |

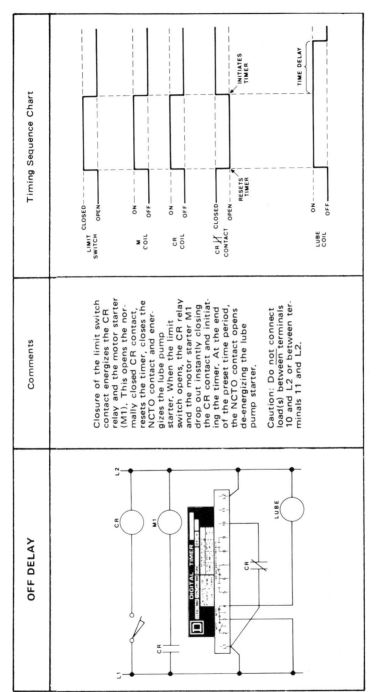

OFF DELAY	Comments	Timing Sequence Chart

Comments

Closure of the limit switch contact energizes the CR relay and the motor starter (M1). This opens the normally closed CR contact, resets the timer, closes the NCTO contact and energizes the lube pump starter. When the limit switch opens, the CR relay and the motor starter M1 drop out instantly closing the CR contact and initiating the timer. At the end of the preset time period, the NCTO contact opens de-energizing the lube pump starter.

Caution: Do not connect load(s) between terminals 10 and L2 or between terminals 11 and L2.

Figure 14-8 Digital solid-state timer. (Courtesy Square D Company.)

OFF DELAY	Comments	Timing Sequence Chart
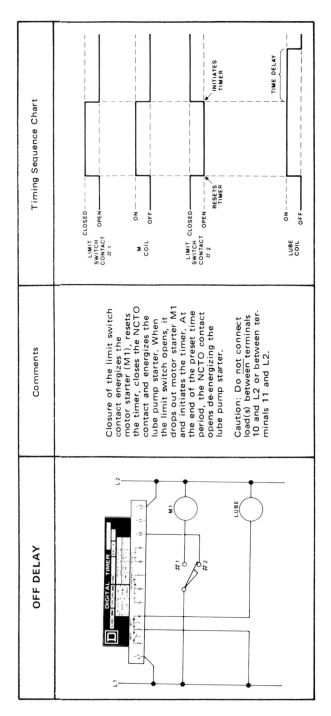	Closure of the limit switch contact energizes the motor starter (M1), resets the timer, closes the NCTO contact and energizes the lube pump starter. When the limit switch opens, it drops out motor starter M1 and initiates the timer. At the end of the preset time period, the NCTO contact opens de-energizing the lube pump starter. Caution: Do not connect load(s) between terminals 10 and L2 or between terminals 11 and L2.	

Figure 14-8 (continued)

MODE OF OPERATION	USER CONNECTION	COMMENTS
HARD CONTACT INPUTS MANUAL PRESET	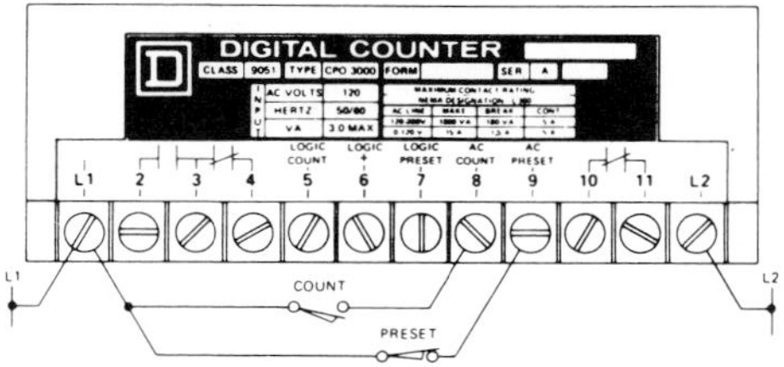	Counter will preset to the number shown on the thumb-wheel switches when the preset switch is opened. Preset switch must be opened to initially preset the counter after L1–L2 power is applied. Preset switch must be closed before counting can commence and must remain closed during counting. Each closure of the count switch advances the counter one count.

HARD CONTACT INPUTS
 	
AUTOMATIC PRESET
	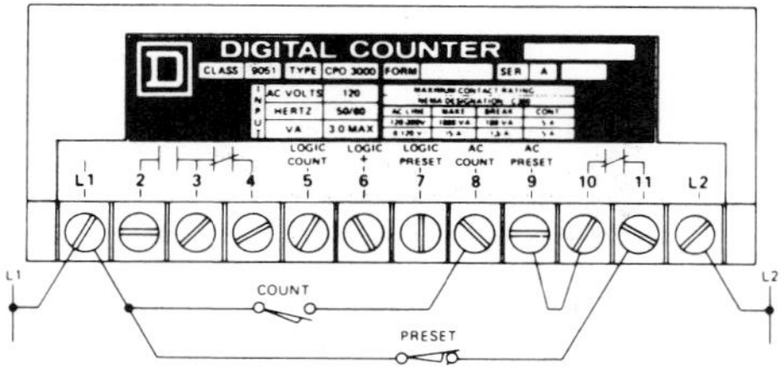	Counter will preset to the number shown on the thumb-wheel switches when the preset switch is opened, or
		
		  1. the count of 000 is reached (when counter is in the down counting mode), or
		
		  2. the count of 999 is reached (when counter is in the up counting mode).
		
		Preset switch must be opened to initially preset the counter after L1–L2 power is applied. Preset switch must be closed before counting can commence and must remain closed during counting.
		
		Each closure of the count switch advances the counter one count.

Figure 14-9 Digital solid-state preset counter. (Courtesy Square D Company.)

MODE OF OPERATION	USER CONNECTION	COMMENTS
NORPAK INPUTS **MANUAL PRESET**	 Typical Anti-Bounce Circuit	Counter will preset to the number shown on the thumbwheel switches when the output of NOR B is a logic "0." Output of NOR B must be a logic "0" to initially preset the counter after L1-L2 power is applied. Output of NOR B must be a logic "1" before counting can commence and must remain a logic "1" during counting. The counter will advance one count each time the output of NOR A switches from a logic "0" to a logic "1." Signals marked with * must be "bounce free." When required, use an anti-bounce circuit as shown to remove "bounce" from an Input Signal.

206

MODE OF OPERATION	USER CONNECTION	COMMENTS

NORPAK INPUTS
 	
AUTOMATIC PRESET
(using form N9)
	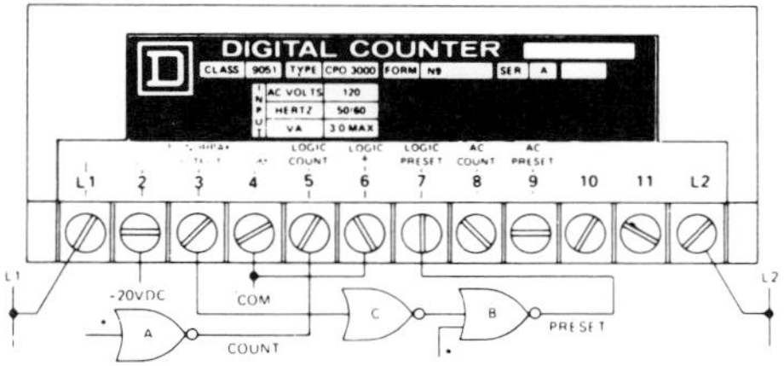 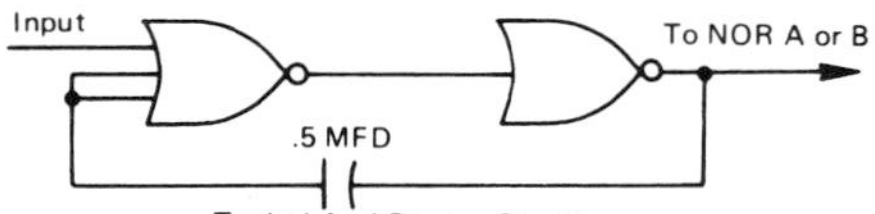
	
	Typical Anti-Bounce Circuit
	
	NOTE: Form N9 may also be used when operating the counter in the Manual Preset mode.
		Counter must first be converted to Form N9 by substituting a Class 9050 Type XA-5 Logic Output Interface Module for the plug-in output relay.
		
		Counter will preset to the number shown on the thumbwheel switches when the output of NOR B is driven to a logic "0" by the * input, or
		
		  1. the count of 000 is reached (when counter is in the down counting mode), or
		
		  2. the count of 999 is reached (when counter is in the up counting mode).
		
		Output of NOR B must be driven to a logic "0" by its * input in order to initially preset the counter after L1-L2 power is applied.
		
		Output of NOR B must be a logic "1" before counting can commence and must remain a logic "1" during counting.
		
		The counter will advance one count each time the output of NOR A switches from a logic "0" to a logic "1."
		
		Signals marked with * must be "bounce free." When required, use an anti-bounce circuit as shown to remove "bounce" from an Input Signal.
		
		Four additional units of NORPAK logic load are available to the user at terminal 3.

30072-252-15-B

Figure 14-9 (continued)

NORPAK is a Registered Trademark of the Square D Company

MODE OF OPERATION	USER CONNECTION	COMMENTS
COS/MOS INPUTS MANUAL PRESET		V_{SS} must be -12 to -15 VDC. V_{DD} is GND. COS/MOS logic elements are CD4049A inverters. Resistors are 1500 ohms, ±10%, 1/2 watt. Counter will preset to the number shown on the thumbwheel switches when the output of inverter B is HIGH (-0.5 volt or more positive). Output of inverter B must be HIGH to initially preset the counter after L1-L2 power is applied. Output of inverter B must be LOW (-10 volts or more negative) before counting can commence and must remain LOW during counting. The counter will advance one count each time the output of inverter A switches from HIGH to LOW. Signals marked with * must be "bounce free."

MODE OF OPERATION	USER CONNECTION	COMMENTS
FILTERED DC VOLTAGE INPUTS MANUAL PRESET	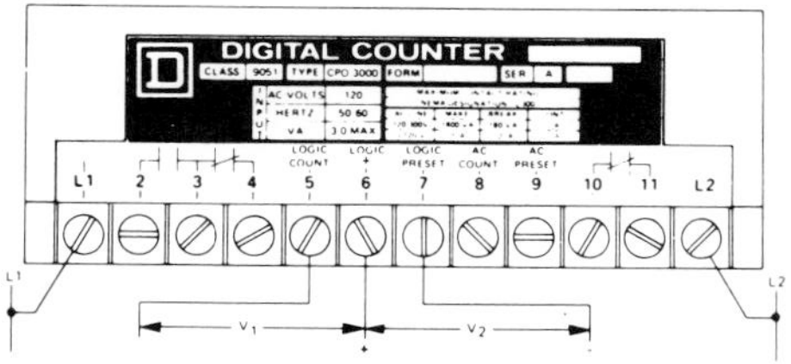	Voltages V_1 and V_2 must be bounce-free, filtered DC voltages with polarity as shown. Counter will preset to the number shown on the thumb-wheel switches when V_2 is 0-0.5 VDC. V_2 must be 0-0.5 VDC to initially preset the counter after L1-L2 power is applied. V_2 must be 10-20 VDC before counting can commence and must remain 10-20 VDC during counting. The counter will advance one count each time V_1 switches from 0-0.5 VDC to 10-20 VDC. Typical sources for V_1 and V_2 are photocells, transistors, and transducers with solid state outputs.
EXTERNAL SWITCH AND LOGIC SIGNAL INPUTS MANUAL PRESET	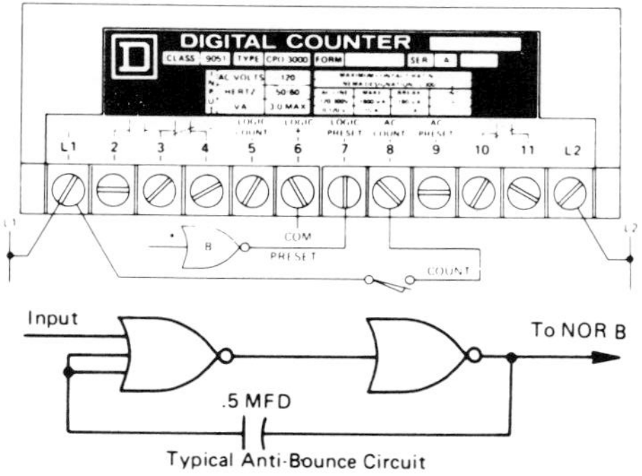	Counter will preset to the number shown on the thumb-wheel switches when the output of NOR B is a logic "0." Output of NOR B must be a logic "0" to initially preset the counter after L1-L2 power is applied. Output of NOR B must be a logic "1" before counting can commence and must remain a logic "1" during counting. Signals marked with * must be "bounce free." When required, use an anti-bounce circuit as shown to remove "bounce" from an Input Signal. Each closure of the count switch advances the counter one count.

Figure 14-10 Digital solid-state preset counter. (Courtesy Square D Company.)

209

MODE OF OPERATION	USER CONNECTION	COMMENTS
EXTERNAL SWITCH AND LOGIC SIGNAL INPUTS AUTOMATIC PRESET	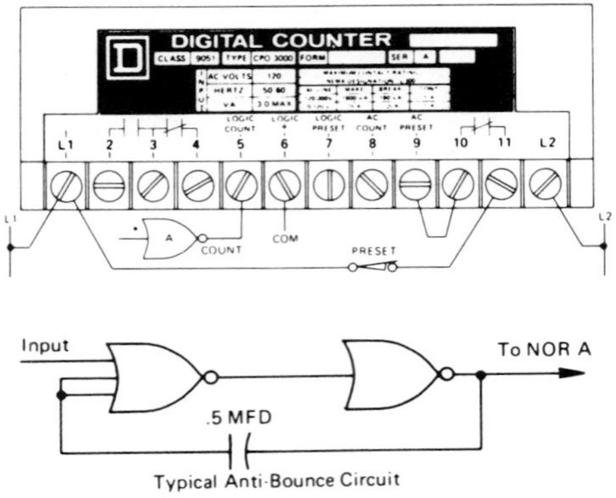 Typical Anti-Bounce Circuit	Counter will preset to the number shown on the thumb-wheel switches when the preset switch is opened, or 1. the count of 000 is reached (when counter is in the down counting mode), or 2. the count of 999 is reached (when counter is in the up counting mode). Preset switch must be opened to initially preset the counter after L1-L2 power is applied. Preset switch must be closed before counting can commence and must remain closed during counting. The counter will advance one count each time the output of NOR A switches from a logic "0" to a logic "1." Signals marked with * must be "bounce free." When required, use an anti-bounce circuit as shown to re-move "bounce" from an Input Signal.

30072-252-15-B

Figure 14-10 (continued)

chapter fifteen

Maintenance of Motor Starters

To ensure long life and trouble-free operation, motor starters must be maintained as for any other equipment. The best approach to good maintenance of motor starters is to use a systematic program of inspection. In general, this means that inspection of motor controls should be made frequently enough to prevent serious trouble. The exact interval between inspections will depend, to a certain extent, on the conditions of the applications involved. However, experience will soon indicate the approximate time interval required.

Technicians who maintain motor-starting equipment must be thoroughly trained for the job and know how to correct any fault that occurs, within a certain reasonable range of expertise. Of course, no maintenance person is expected to be able to completely overhaul any piece of equipment that breaks down. This is often left to specialists in the field who are trained for specific types of equipment. However, maintenance personnel should have a good working knowledge of how motor starters operate and how to correct at least any minor faults that occur from time to time.

Good maintenance personnel, coupled with a good preventive maintenance program, will almost certainly keep motor controllers operating in good condition the majority of the time. In general, this program will consist of continuing inspection of equipment, the report and recording of the condition of the equipment, and equipment repair.

Although the exact details of a preventive maintenance program will vary with the conditions of the applications, the following general points should apply to almost all conditions:

1. Everything possible must be done for the safety of personnel.
2. Initial installation should be tested and proved satisfactory before it is accepted. Apparatus should be easily accessible for inspection and repair.
3. An adequate supply of correct renewal parts must be available.
4. Enclosures should be chosen with respect to the operating conditions.
5. Controllers should be kept clean and dry.
6. Contacts that are worn very thin or badly burned and pitted should be replaced. Replacement should be by pairs, maintaining correct contact pressures.
7. Contacts should be kept clean. Contact shape should not be changed by rough filing or grinding.
8. Contacts and all connections should be kept tight.
9. Contactor or relay bearings should not be oiled but should be kept clean and with no friction in the moving parts.
10. Coils should be operated at rated voltage. Both overvoltage and undervoltage conditions are undesirable.
11. Arc-rupturing parts should be kept in good condition and in correct operating positions.
12. Frayed and worn shunts should be replaced.
13. All dashpots should be kept clean and all oil dashpots should have the correct oil in them.
14. Conditions that cause excessive temperatures should be corrected. Temperatures should be measured if in doubt about overheating.
15. Undesirable grounds on all circuits should be watched for and eliminated.

Taking care of these items, one by one, is warranted at this point to obtain a better understanding of the techniques involved. All are important to better motor control operation and safety.

Basically, safety is accomplished by enclosing, guarding, or remotely locating apparatus to avoid personal contact, by disconnecting all power before touching or repairing equipment, and by following recognized safety rules.

Before putting any electric motor controller into operation, all installation work and wiring connections should be thoroughly checked to ensure that the work has been performed according to all codes and in a workmanlike manner. All equipment and parts should have ample capacity for the loads they carry. Tests should be made to verify this adequacy.

For inspection and repair work, all parts should be made as accessible as possible. It should be possible to renew contacts, coils, springs, and other important parts quickly and with few tools. Installation should be arranged so that all units are accessible for maintenance work.

In most instances, when a motor or group of motors are shut down, the result can be costly in lost production time. Therefore, an adequate supply of renewal parts should be kept on hand at all times to prevent unnecessary downtime. Renewal parts should be obtained from the manufacturer of the original equipment to make certain that the parts are correctly made of the proper materials. Substitutes would have to be thoroughly checked for dimensional accuracy and correct materials.

Most controllers are mounted in some type of metal enclosure. This is done to protect personnel from live parts and to protect the controller from mechanical damage, corrosive conditions, and unauthorized tampering. The most common enclosure is made of sheet metal that encloses all live parts. It may be ventilated or nonventilated.

In areas where dust, such as coal dust, cement dust, etc., is present dust-tight enclosures will reduce maintenance of controller parts. These enclosures require gaskets and are made such that dust and dirt cannot enter. Weather-resistant, drip-tight, watertight, and submersible-type enclosures are necessary when there are corresponding service conditions.

For hazardous locations such as mines, refineries, cleaning plants, or wherever explosive atmospheres are present, enclosures are made of heavy construction. They are designed such that they will withstand explosions of gases within the enclosure, without damage to the enclosure and without permitting any sparks or flames to emerge to cause a fire or an explosion.

Oil-immersed controllers are used to prevent explosions in dangerous locations. Where acid fumes or other highly corrosive atmospheres exist, maintenance work will be minimized if the controller parts are immersed in oil and the enclosure is protected by a suitable resistant finish.

Very few industrial controllers operate in clean locations. Oil and moisture are often present as liquids or as vapors in the air. Dust, lint, or other materials are present as the natural result of operating conditions. These materials, separately or in combination, create maintenance work. They reduce the insulating distance across otherwise clean and dry surfaces. They collect dust and dirt that may cause sluggish mechanical or electrical action.

Accumulations of dust and dirt should regularly be removed either by vacuum or by blowing with compressed air. Excessive air pressures should be avoided because sharp, small particles may be driven into some insulating materials. Special attention may be required to remove metallic dust with magnetic properties, which readily collects and adheres to the magnetized parts of the controller. Dirt, oil, and moisture are most easily removed by wiping the surfaces with cloths and suitable solvents.

Moisture due to condensation may collect within an enclosure. Drainage holes are rarely acceptable to relieve this condition. Heaters are most often used to prevent moisture by condensation. Heaters are most essential when the controller is idle. When in operation the coils and resistors within the enclosure will usually provide enough heat to prevent condensation.

Each time motor controller contacts open and close they are subject to mechanical wear and electrical burning. The contacts close with a rolling movement combined with a wiping action which causes wearing of the contact material. Contact parts may therefore require considerable maintenance, depending on the operating conditions. The actual mechanical wear of the contacts may be more serious than the electrical burning caused by arcing.

As contacts wear, the material in them gradually disappears as a result of mechanical wear and electrical burning. During the wearing process the contact pressures decrease. This affects the current-carrying ability of the contacts, which if allowed to go too low will cause the contacts to overheat. A small contact with suitable pressure will carry current with less heating than will a large contact with little or no pressure. Reasonable provisions are made in the original design for the wearing of contacts, but replacements will eventually be necessary. If contact springs have been overheated, they may be unable to provide sufficient contact pressure because the material has been weakened by overheating. Contact pressures should be checked and maintained within suitable limits. Both moving and stationary contacts should always be replaced.

Since controller contacts operate in pairs, a single contact should never be replaced; a new contact should never be allowed to operate in conjunction with an old contact. Because of the wearing of contact surfaces, the probability is very high that a mixture of old and new parts will operate poorly. The small additional expense of replacement in pairs will be repaid many times over in operating life.

Contacts are generally made of copper or silver. Silver contacts are generally used on the small current-carrying contacts of relays, electrical interlocks, pushbuttons, thermostats, pressure switches, and similar devices. The remainder are made of copper, and large contacts for heavy currents are almost always made of copper. Contacts should be kept clean. This is especially true of copper contacts because the copper oxide that soon appears on clean copper is not a good electrical conductor. It increases contact resistance and is often the cause of serious heating of contacts. When a contact is renewed it is important to clean it if it is discolored. The surface against which it is mounted should also be cleaned.

The slight rubbing action and burning that occur during normal good operation will generally keep contact surfaces clean enough for good service. Copper contacts that seldom open or close, however, will readily accumulate a thin discolored surface that may cause heating. This is not true of silver contacts. The dense discoloration that soon appears on clean silver is a relatively good electrical conductor. It is not necessary to keep silver contacts clean except for the sake of appearance.

When excessive currents are closed or opened, or when contact motion is sluggish, contact surfaces may be severely burned. If this burning causes deep pits or craters or a very rough surface, both the stationary and moving contacts should be renewed. It is not, however, essential, or even desirable, to have entirely smooth contact surfaces. Slightly roughened surfaces that appear during normal good operation, if clean, provide a better contact area than do smooth surfaces. Contacts with surfaces comparable to very coarse sandpaper may therefore be considered to be in good condition.

Dirty or excessively rough contacts should be cleaned and smoothed with a fine file, but care should be used to maintain the true surface shape or contour of the original contact. The designer has spent much time and effort to determine the best contact shape. Changing the original shape by careless filing will leave high points or edges that may overheat. Emery paper should not be used to clean contacts since it is an electrical conductor. Furthermore, some particles become embedded in the contact surfaces and will cause unnecessary wear.

Any loose electrical connection will eventually cause trouble. An open circuit or an unreliable circuit may cause considerable lost time and production because they are often very difficult to find. A loose connection can cause a poor contact of high resistance. The discoloration or copper oxide buildup increases the resistance of the contact surface. The higher resistance causes more heating and the increased temperature causes more oxidation and higher resistance. The effect is always cumulative and the heating increases until the parts overheat, deteriorate, or burn. Other loose connections cause similar heating and on thermally operated devices, such as a heater of a thermal overload relay, may cause the relay to trip and stop a motor when the motor is not overloaded.

The bolts or fastening devices that hold contacts in place should always be tight. Normal expansion and contraction of metals due to temperature changes or excessive vibration will cause bolts or nuts to become loose. Frequent checking for loose contacts is therefore advisable.

Very few contacts close without some bounce or rebound. This is due to the reaction of the contact springs as they are compressed to provide the final contact pressure. When the contacts bounce, they separate. At this time the contacts are carrying current, and even though the separation is very small, an arc is created. This arc, if severe, may cause sharp projections of burned or roughened contact surfaces to overheat and may weld or "freeze" the contact surfaces together. Under such conditions, the contacts may not open when next expected to do so. Other causes of contact welding are excessive currents when contacts close or open, insufficient contact pressure, sluggish operation either when closing or opening, and momentary closure of contacts without much, or any, pressure applied. Well-designed contacts, properly applied, reduce this hazard to a minimum.

Coils provide the electromagnetic pull that causes the contacts of relays and contactor to open and close. Series coils generally carry heavy currents and have relatively few turns of rather heavy copper. Shunt coils have many turns of insulated wire. They are generally impregnated in a vacuum or under pressure with insulating compounds and are covered with insulating tape or other material. Impregnating compounds produce a firm but resilient binding material that prevents cracks when temperature changes occur. The impregnation process eliminates air pockets within the winding and makes the coil a solid mass that is better able to radiate heat and is less subject to mechanical injury.

Shunt coils for ac devices are designed to close the devices at 85% of rated voltage. Coils for dc devices will close them at 80% of normal voltage. Any coil is expected to withstand 110% of rated voltage without damage.

A coil with an open circuit will not operate a contactor or relay. The questionable coil should be immediately replaced by one that is known to be satisfactory. The questionable coil should then be checked for open circuits.

If some turns of a coil become short-circuited, the resistance of the coil will be reduced. More current then passes through the coil. The increased current will cause higher operating coil temperatures and will often cause coil burnout.

Coils should be operated at the rated voltage. Overvoltage on coils causes them to operate at a higher temperature that shortens coil life. Overvoltage also operates the contactor or relay with unnecessary force and causes more mechanical wear and bounce when closing.

Undervoltage on coils causes contactors and relays to operate sluggishly. The contact tips may touch but the coils may be unable to completely close the contacts against the contact spring pressure. Under these conditions, the contact pressure is below normal and the contacts may overheat and weld together.

Due to the magnetic air gap and the characteristics of ac circuits, the coil current of an ac contactor is much higher while the contactor is closing and the air gap is larger than after the contactor has closed and the air gap is zero. Since the closing time is short, ac coils are designed to withstand the closed conditions. They will soon overheat if the unit is blocked or the voltage is too low to close the contact and the coil remains energized with a large air gap in the magnetic circuit. Freely operating parts will avoid such coil burnouts. Dc coils are not subject to these conditions because the coil currents do not vary with the air gap.

When controller contacts are expected to open circuits carrying currents that are difficult to interrupt, they are equipped with arc-rupturing parts. These parts generally surround the contacts and must be made such that they are easily moved out of position or removed entirely in order to inspect and replace both moving and stationary contacts. To be effective, the arc-rupturing parts must be in a definite position with respect to the contacts. Hence arc-rupturing parts should always be returned to the proper position if removed for any reason.

On dc starters the arc shields are generally made of a molded material. The arc shield should always be down so that the arc is broken within the field of the blowout coil; otherwise, the shield will not give satisfactory results. Arc shields should be renewed before the molded material is burned away sufficiently to expose the metal parts to the arc.

One of the principal and interesting differences between the conventional dc contactor with its magnetic blowout coil and the modern small ac contactor is the deionizing arc quencher. Deionizing arc quenching action confines, divides, and extinguishes the arc almost the instant the circuit is broken. The usual flash and scattering of flame outside the arc box are thus done away with.

Deionizing action not only stretches the arc but also confines and quenches it.

As the specially shaped contacts separate, a magnetic reaction occurs that causes the stretching arc to rise immediately into the deionizing grids, where it is sliced into a series of arcs. At the next zero point on the current cycle, the air adjacent to each grid is deionized instantly, the voltage per grid is insufficient to reestablish the current flow, and the arc is out.

The arc quencher is assembled over the contacts in much the same manner as the magnetic blowout coils and arc shields on the conventional dc contactor and it is easily removed for inspection of the contacts. Pitting and burning of contacts and the arc box are greatly reduced and longer, trouble-free life is provided with the deionizing system.

The fine strands of flexible shunts sometimes break where the shunt bends. The unbroken strands must then carry the entire current. If several strands break, the unbroken ones become overloaded. They then overheat and eventually burn. Frayed shunts should therefore be promptly replaced by new ones.

Air or oil dashpots are used to retard motion. They are machined to close clearances and must be kept clean and free to move. The proper amount and type of oil should always be kept in oil dashpots. Since the viscosity of oil changes with temperature changes, substitute oils should not be used.

Overheated parts always indicate trouble. However, it is often difficult to know when temperatures are excessive. Resistors are operated safely at 360°C rise above ambient temperature, but insulated coils are generally restricted to 85°C above ambient. Solid copper contacts are limited to a rise of 65°C and copper buswork to 50°C.

Barring the presence of gases, acids, or alkalis, much-discolored copper parts have been or are too hot. When in doubt, temperatures should be measured by thermometer or other means. Touch of the hand is unreliable because safe operating temperatures of many electrical parts are unbearably hot to the hand. It is best to know what the permissible temperatures are and then measure them.

Grounds are both useful and undesirable. Desirable grounds are rather easily maintained because they require only good contact connections. An unexpected ground is a serious personal hazard. Constant vigilance is required to prevent and eliminate undesirable grounds. They cause operating trouble and erratic and dangerous operating circuits. Because of grounds, motors may start unexpectedly, motors may not stop when they should, and overload and other protective features may be made ineffective.

Grounds often occur in pushbutton boxes or similar confined spaces where stray strands of wire may make contact at incorrect places. They occur when wires become chafed due to vibration on rough edges such as conduit entrances. Conduits sometimes become wet from condensation or other reasons and the insulation on the wires becomes water soaked and of low insulating value. Conduits should therefore be installed so that moisture within them will always drain away. It may sometimes be necessary to remove the wires, clean the conduits, and install new wiring. Clean and dry conditions always reduce maintenance.

Table 15-1 lists problems encountered with motor controls, together with their causes and remedies. This table is of a general nature and covers only the main causes of problems.

Misapplication of a device can be a cause of serious trouble and is a major cause of motor control trouble and should always be found quickly and corrected.

TABLE 15-1   Troubleshooting and Maintenance of Square D Motor Control Equipment

Danger: Hazard of electrical shock or burn. Be sure to turn off power supplying this equipment before working on it.

MOTOR CONTROL TROUBLE—REMEDY TABLE

The following table lists troubles encountered with motor control, their causes and remedies. This table is of a general nature and covers only the main causes of trouble.

Misapplication of a device can be a cause of serious trouble; however, rather than list this cause repeatedly it should be noted here that *misapplication is a major cause of motor control trouble and should always be questioned when a device is not functioning properly.*

Actual physical damage or broken parts can usually be found quickly and replaced. Damage due to water or flood conditions requires special treatment.

Trouble	Cause	Remedy
Contacts		
Contact chatter (*also see* "Noisy magnet")	1. Poor contact in control circuit.	1. Replace the contact device or use holding circuit interlock (three-wire control).
	2. Low voltage.	2. Check coil terminal voltage and voltage dips during starting.
Welding or freezing	1. Abnormal inrush of current.	1. Check for grounds, shorts, or excessive motor load current, or use larger contactor.
	2. Rapid jogging.	2. Install larger device rated for jogging service.
	3. Insufficient tip pressure.	3. Replace contacts and springs, check contact carrier for deformation or damage.
	4. Low voltage preventing magnet from sealing.	4. Check coil terminal voltage and voltage dips during starting.
	5. Foreign matter preventing contacts from closing.	5. Clean contacts with Freon. Contacts, starters, and

(continued)

TABLE 15-1   (*Continued*)

Trouble	Cause	Remedy
		control accessories used with very small current or low voltage, should be cleaned with Freon.
	6. Short circuit or ground fault.	6. Remove fault and check to be sure fuse or breaker size is correct.
Short tip life or overheating of tips	1. Filing or dressing.	1. Do not file silver tips. Rough spots or discoloration will not harm tips or impair their efficiency.
	2. Interrupting excessively high currents.	2. Install larger device or check for grounds, shorts, or excessive motor currents.
	3. Excessive jogging.	3. Install larger device rated for jogging service.
	4. Weak tip pressure.	4. Replace contacts and springs, check contact carrier for deformation or damage.
	5. Dirt or foreign matter on contact surface.	5. Clean contacts with Freon. Take steps to reduce entry of foreign matter into enclosure.
	6. Short circuits or ground fault.	6. Remove fault and check to be sure fuse or breaker size is correct.
	7. Loose connection in power circuit.	7. Clear and tighten.
	8. Sustained overload.	8. Check for excessive motor load current or install larger device.
Coils Open circuit	1. Mechanical damage.	1. Handle and store coils carefully.
Overheated coil	1. Overvoltage or high ambient temperature.	1. Check coil terminal voltage, which should not exceed 110% of coil rating.
	2. Incorrect coil.	2. Install correct coil.
	3. Shorted turns caused by mechanical damage or corrosion.	3. Replace coil.
	4. Undervoltage, failure of magnet to seal in.	4. Check coil terminal voltage, which should be at least 85% of coil rating.

(*continued*)

TABLE 15-1 *(Continued)*

Trouble	Cause	Remedy
	5. Dirt or rust on pole faces.	5. Clean pole faces.
	6. Mechanical obstruction.	6. *With power off,* check for free movement of contact and armature assembly.
Overload relays Tripping	1. Sustained overload.	1. Check for excessive motor currents or current unbalance, and correct cause.
	2. Loose or corroded connection in power circuit.	2. Clean and tighten.
	3. Incorrect thermal units.	3. Thermal units should be replaced with correct size for the application conditions.
	4. Excessive coil voltage.	4. Voltage should not exceed 110% of coil rating.
Failure to trip	1. Incorrect thermal units.	1. Check thermal unit selection table. Apply proper thermal units.
	2. Mechanical binding, dirt, corrosion, etc.	2. Replace relay and thermal units.
	3. Relay previously damaged by short circuit.	3. Replace relay and thermal units.
	4. Relay contact welded or not in series with contactor coil.	4. Check circuit for a fault and correct condition. Replace contact or entire relay as necessary.
Magnetic and mechanical parts Noisy magnet	1. Broken shading coil.	1. Replace magnet and armature.
	2. Dirt or rust on magnet faces.	2. Clean.
	3. Low voltage.	3. Check coil terminal voltage and voltage dips during starting.
Failure to pick up and seal	1. No control voltage.	1. Check control circuit for loose connection or poor continuity of contacts.
	2. Low voltage.	2. Check coil terminal voltage and voltage dips during starting.
	3. Mechanical obstruction.	3. *With power off,* check for free movement of contact and armature assembly.

(continued)

TABLE 15-1   (*Continued*)

Trouble	Cause	Remedy
	4. Coil open or overheated.	4. Replace.
	5. Wrong coil.	5. Replace.
Failure to drop out	1. Gummy substance on pole faces.	1. Clean pole faces.
	2. Voltage not removed.	2. Check coil terminal voltage and control circuit.
	3. Worn or corroded parts causing binding.	3. Replace parts.
	4. Residual magnetism due to lack of air gap in magnet path.	4. Replace magnet and armature.
	5. Contacts welded.	5. *See* ''Contacts—Welding or freezing''
Pneumatic timers Erratic timing	1. Foreign matter in valve.	1. Replace complete timing head or return timer to factory for repair and adjustment.
Contacts do not operate	1. Maladjustment of actuating screw.	1. Adjust per instruction in service bulletin.
	2. Worn or broken parts in snap switch.	2. Replace snap switch.
Limit switches Broken parts	1. Overtravel of actuator.	1. Use resilient actuator or operate within tolerances of the device.
Manual starters Failure to reset	1. Latching mechanism worn or broken.	1. Replace starter.

Source: Courtesy Square D Company.

Damage due to water or flood conditions requires special treatment. See Chapter 16 for recommendations for renovating water-soaked motor controls.

Table 15-2 lists good maintenance practices as well as troubleshooting hints and remedies. Tables 15-1 and 15-2 combined should take care of the majority of motor control problems, provided that maintenance personnel are trained in this field and understand the use of the tables.

TABLE 15-2   Motor Starter Check Chart

Trouble	Cause	What to Do
Contactor or relay does not close	No supply voltage.	Check fuses and disconnect switch.
	Low voltage.	Check power supply. Wire size may be too small.
	Open-circuited coil.	Replace.
	Pushbutton, interlock, or relay contact not making.	Adjust for correct movement, ease of operation, and proper contact pressure.
	Loose connections or broken wire.	Check circuit with flashlight (turn power off first).
	Pushbutton not connected correctly.	Check connections with wiring diagram.
	Overload relay contact open.	Reset relay.
	Damaged, worn, or poorly adjusted mechanical parts.	Clean and adjust mechanically. Align bearings and free the movement. Repair or replace worn or damaged parts.
Contactor or relay does not open	Pushbutton not connected correctly.	Check connections with wiring diagram.
	Shim in magnetic circuit[a] worn, allowing residual magnetism to hold armature closed.	Replace shim.
	Pushbutton, interlock, or relay contact not opening coil circuit.	Adjust for correct movement, ease of operation, and proper opening.
	"Sneak" circuits.	Check apparatus and wiring for insulation failure.
	Contacts weld shut.	See "Excessive corrosion of contacts."
	Damaged, worn, or poorly adjusted mechanical parts.	Clean and adjust mechanically. Align bearings and free the movement. Repair or replace worn or damaged parts.
Excessive corrosion of contacts—contacts weld shut—contacts overheat	Insufficient contact spring pressure causing contacts to overheat or draw an arc on closing.	Adjust, increasing contact pressure. Replace spring or worn contacts if necessary.
	Rough contact surface causing current to be carried by too small an area.	Dress up contacts with fine file. Replace if badly worn.
	Abnormal operating conditions.	Check rating against load. If conditions are too severe

(*continued*)

TABLE 15-2   (*Continued*)

Trouble	Cause	What to Do
	Chattering of contacts as a result of vibrations outside of controller cabinet.	for open-type contactors, replace with oil-immersed or dust-tight equipment. Instruct operator in proper manipulation of manually operated device. Check control switch contact pressure and replace spring if it does not give rated pressure. Tighten all connections. If this does not help, mount or move control, so that vibrations are decreased.
	Sluggish operation.	Clean and adjust mechanically. Align bearings and free movement.
Arc lingers across contacts	If blowout is series it may be shorted. If blowout is shunt it may be open circuited.	Look at wiring diagram and see type of blowout, then check to see if circuit through blowout is all right.
	Ineffective blowout coil.	Check rating and if improperly applied replace with correct coil. Check polarity and reverse coil if necessary.
	If no blowout is used, note travel of contacts.	Increasing travel of contacts will increase rupturing capacity.
	If used, arc box might be left off or not in correct position.	See that arc box is fully in place.
	Overload.	Check rating against load.
	Creepage or voltage breakdown over or through arc box wall.	Clean—dry out in oven, or replace.
Noisy ac magnet	Improper seating of the armature.	Adjust mechanical parts and clean pole faces.
	Broken shading coil.	Replace.
	Low voltage.	Check power supply. Wire size may be too small.
Abnormally short coil life	High voltage.	Check supply voltage against rating of controller.
	Gap in magnetic circuit.[b]	Check travel of armature and

(*continued*)

TABLE 15-2 (*Continued*)

Trouble	Cause	What to Do
	Too high an ambient temperature.	adjust so that magnetic circuit is completed. Clean pole faces. Check rating of controller against ambient temperature. Get coil of higher ambient rating from manufacturer if necessary.
Panel and apparatus burned by heat from starting resistor.	Motor being started too frequently.	Use resistor of higher rating.

*a*Dc only.

*b*Ac only.

Source: Courtesy Westinghouse.

chapter sixteen

Water-Soaked Motor Controls

GENERAL RECOMMENDATIONS

Motor controls will sometimes become water soaked because of an activated sprinkler system, a fire in part of the motor control area where water is used to extinguish the blaze, or perhaps a flash flood in the area.

When such a condition exists, all equipment should be hosed down and scrubbed with a stiff brush. Water pressure should be limited to 25 pounds per square inch maximum on insulating parts. Water temperature should not exceed 90°C (194°F) maximum. Freon or any approved solvent, such as trichloroethylene or AWA 1,1,1, should be used for a final cleaning of insulating surfaces, contact tips, and magnet pole faces. Steam cleaning should not be used except on cabinets and enclosures from which the interiors have been removed.

Drying in the sun or in a warm, dry room will take about 2 to 4 days. If heat is used for drying, it should be limited to about 212°F, preferably starting at about 150°F or lower and gradually building up to 212°F. Plaster drying equipment, electric heaters, or baking ovens may be used. Approximately 24 hours of drying by this method should generally be sufficient. For equipment operating at voltages above 250 V, greater caution in drying and cleaning should be taken, because of the greater susceptibility to insulation breakdown.

As a final check, test insulation resistance with a megger or ohmmeter. A 500-V or 1000-V megger is recommended; higher voltages are not suitable for testing controls rated at 600 V. One megohm is recommended as minimum resistance, but equipment may operate satisfactorily with less. In operation, normal drying will cause this resistance value to increase. Restore power to one device at a time and watch for leakage, splitting, or smoking. If all equipment passes this test, load may then be restored, but still watch for spitting or smoking at contacts. Check to be sure that no open-circuit conditions exist.

SPECIFIC RECOMMENDATIONS

Coils and transformers.   Wash with clean hot water and dry by any of the methods listed above or by applying a low voltage (about 10% or rated voltage) to the coil for 2 to 3 hours. The magnet armature should be in the open position or the coil should be removed from the magnet. For drying dc coils by this method, a resistor in parallel with the coil is recommended to provide a discharge path for the kick voltage induced when the circuit is opened. Special care should be taken to dry out dc coils, since in operation, the inductive voltage may be several times larger than the normal voltage. Check with a megger for insulation resistance (1 megohm).

Magnets.   Wash, clean, and dry with a cloth or an air hose. Clean magnet pole faces with Freon to remove oil film and surface dirt, which might cause the armature to stick.

Contact surfaces.   Clean tips with Freon. Filing the tips is not recommended, but it may be done if the surface is very difficult to clean. Do not use sandpaper or emery cloth, which would deposit grit on the tips. Small contacts can be "filed" by rubbing kraft paper over the surface. Particular care should be taken with small contacts such as electrical interlocks, overload relays, and pushbuttons.

Insulating parts.   All surfaces of contact blocks, arc chamber covers, movable contact carriers, overload relay blocks, and so on, should be cleaned with Freon, particularly between poles, and dried carefully. Surface drying is generally sufficient for glazed porcelain, Melamine, and nonabsorbent plastics. Unglazed porcelain, Bakelite, and Rostone absorb water and should be baked to dry thoroughly. Check with a megger from pole to pole and from each pole to ground for 1 megohm minimum insulation resistance.

Thermal overload relays.   Remove melting-alloy thermal units, and inspect mating surfaces of the thermal units and overload relay block. Clean with Freon where necessary. Replace thermal units and tighten fastening screws. Tight screws and clean electrical connections are essential in preventing nuisance tripping. Bimetallic overload relays should be inspected and cleaned in the same way.

To inspect operation of the overload relay contact, use a continuity checker and trip the relay. Trip the melting-alloy relay by lifting the pawl from the thermal unit. A bimetallic relay is tripped by removing a heater element and lightly pressing on the lower end of the bimetallic strip with a screwdriver. If there is a mechanical problem, replace the entire overload relay.

Snap switches. Snap switches should generally be replaced as soon as possible. For temporary use, shake out any water trapped inside, blow the switch out with an air hose, and bake to dry. Check the circuit continuity. Dirt inside the case may prevent contacts from closing properly and cause excessive wear on the toggle mechanism. It is generally impractical to attempt to open the case to clean it out.

Pneumatic timers. Timer heads should not be taken apart. The timer should be replaced or returned to the factory for repair. For temporary use, dry the timer thoroughly and treat the coil, magnet, and snap switch as recommended. Check the timing operation. The timer should be suitable for limited use, but it will probably not be very accurate because of the dirt in the valves.

CONTACT TIPS

Figure 16-1 shows contact tips in three conditions: new, still serviceable, and needing replacement. There are two types of contact wear: electrical and mechanical. The majority of wear to contact tips is the result of electrical wear. The mechanical wear is insignificant and requires no further mention.

Arcing causes electrical wear by eroding the contacts. During arcing, a small part of each contact melts, vaporizes, and is blown away.

When a device is new, contacts are smooth and have a uniform silver color. With use, the contacts of the device become pitted and the color may change to blue, brown, and black. These colors result from the normal formation of metal oxide on the contact's surface and are not detrimental to contact life and performance. Therefore, do not file contacts because filing only shortens contact life and it may cause welding.

The contacts should be replaced under the following conditions:

1. *Insufficient contact material:* When less than $\frac{1}{64}$ in. remains, replace the contacts.
2. *Irregular surface wear:* This wear is normal, but if a corner of the contact material is worn away and a contact may mate with the opposing contact support member, the contacts should be replaced. This condition can result in contact welding.
3. *Pitting:* Under normal wear, contact pitting should be uniform. This condition occurs during arcing, as described above. The contacts should be replaced if pitting becomes excessive and little contact material remains.
4. *Curling or contact surface:* This condition results when severe service produces high contact temperatures that cause the contact material to separate from the contact support member.

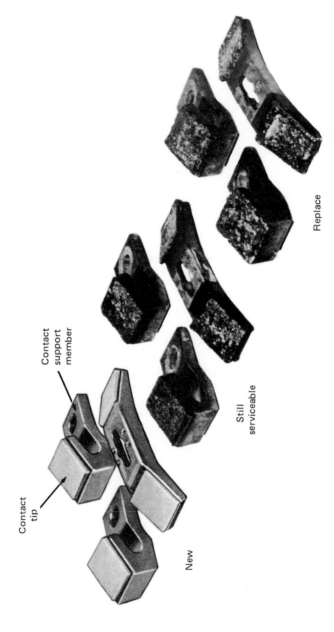

Contact
support
member

Contact
tip

New

Still
serviceable

Replace

Figure 16-1 Three conditions of motor control contacts. (Courtesy Square D Company.)

The measurement procedure for checking the contact tip material requires a continuity checker and a $\frac{1}{31}$-in. feeler gauge. The procedure is as follows:

1. Place the feeler gauge between the armature and the magnet frame with the armature held tightly against the magnet frame.
2. Check the continuity of each phase.

If there is continuity in all phases, the contacts are in good condition; if not, all contacts (both stationary and movable) should be replaced. Even though the contacts pass condition 1, any of the other conditions would necessitate replacement of the contacts. Contacts should be replaced only when necessary; too frequent replacement is a waste of money and natural resources.

chapter seventeen

Methods of Deceleration

There are many times during use when a motor must be controlled in ways other than by normal starting and stopping. Some of the most used operations are plugging, jogging, and braking.

PLUGGING

Plugging is an operation in which the connection to a motor is quickly reversed for an instant, causing reverse torque on the motor to bring it to an abrupt stop. This technique may be used either for quick stopping or for rapid reversing of motors.

A special switch known as a *plugging switch* or a *zero-speed switch* is used to provide plugging. A plugging switch is a centrifugally operated device mounted on the motor shaft where it can "sense" when motor rotation has ceased and will then open the reverse contactor. Such switches are often installed on machine tools that are required to come to an abrupt stop at some point in their cycle of operation to prevent damage to the machine or the work itself. Plugging is also used in processes where machines must come to a complete stop before the next phase of operation is begun; the reduced stopping time saves production time.

In operation, the shaft of the plugging switch (Fig. 17-1) is mechanically connected to the motor shaft or possibly to a shaft on the driven machine, but it is usually best to avoid the latter connection if at all possible. When the switch is mounted in this manner, the motion of the motor is transmitted to the switch contacts by a centrifugal mechanism or a magnetic induction arrangement within the switch. The contacts are wired to the reversing starter, which controls the motor.

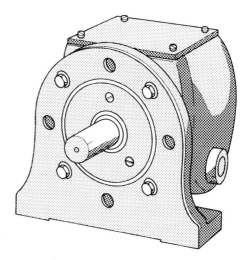

Figure 17-1 Typical   motor   plugging
switch.

Acting as a link between the motor and reversing starter, the switch allows the
starter to apply just enough ''reverse'' power to bring the motor to a quick stop.

JOGGING

The jogging circuit shown in Fig. 17-2 is used primarily when machines must be
operated momentarily for inching, such as in a machine tool set up for maintenance.
The jog circuit energizes the starter only when the jog button is depressed, thereby
giving the machine operator instantaneous control of the motor drive. When the jog
button is depressed, the control relay is bypassed, and the main contactor coil is
energized solely through the jog button; when the jog button is released, the
contactor coil releases immediately. Pushing the start button closes the control
relay, and the relay is held in by its own normally open contacts. The main contactor coil is in turn closed by another set of normally open contacts on the control
relay and is held in the ''on'' position.

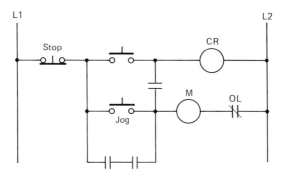

Figure 17-2 Schematic diagram of a jogging motor control circuit.

ELECTRIC BRAKES

Electric motor brakes consist of a linkage of moving parts and a friction member. The brakes are applied to hold a load and/or retard the normal rotation of some machines. Different types are available depending on the power supply, ratings, and the application requirements.

Type SA Brakes

Type SA brakes have a dc clapper-type magnet and are designed so that when the magnet is energized, the shoe will clear the wheel, and when deenergized, the shoes are pressed against the wheel by means of compression springs. The force of the compression springs produces equal pressure of the shoes against the wheel, and when the magnet is energized, each shoe is automatically moved away from the wheel by an equal amount.

The type SA brake is designed to be self-adjusting so that no adjustment is necessary to compensate for lining wear. The travel of the magnet is almost constant, so that the current requirements for releasing the brake do not increase as the lining wears but remain uniform from new lining to worn lining.

The compression springs producing the shoe pressure have a large amount of compression, so that the variation in torque from new lining to worn-out lining is less than 10%.

The brake is provided with a means of releasing the brake by hand when necessary for removing the brake shoes or wheel. The wheel can be removed by loosening the shoe bolts and lifting the wheel up without disturbing any adjustment.

The coil and clutch mechanism are mounted in a weatherproof, dust-proof housing which prevents magnetic dust from entering the housing and sticking to the mating faces of the magnet. The torque rating is marked on an indicator pin and can be changed by turning the main-spring adjusting bushing.

Proper shoe clearance is obtained by setting an adjusting bushing and locking it in position with a locknut. Once made, this adjustment should not need to be changed.

Operation

Refer to Figs. 17-3 and 17-4 and note the following:

(a) Compression springs 1 and 2 (Fig. 17-3) exert a downward force on spindle 3, which, acting on the end of lever 4, applies the shoes against the wheel with equal force. The amount of the spring force is adjusted by raising or lowering upper spring seat 5 by means of adjusting bushing 6. Turning this bushing counterclockwise increases the spring compression, and turning clockwise decreases it. Indicating pin 7 shows the torque rating for each frame.

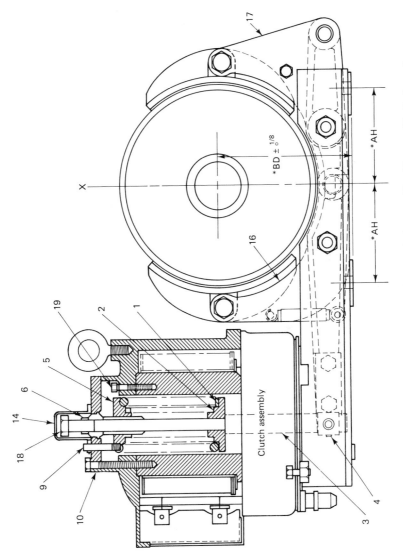

Figure 17-3 Sectional view of type SA electric motor brake assembly. (Courtesy Westinghouse.)

233

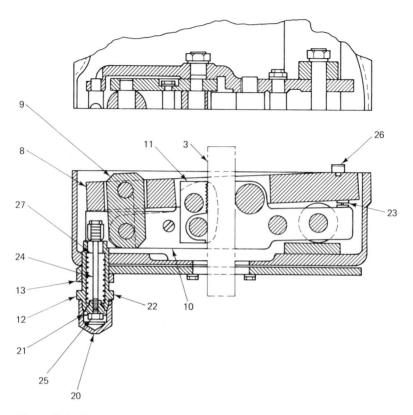

Figure 17-4 Typical clutch assembly on an electric motor brake. (Courtesy Westinghouse.)

(b) When the coil is energized, clapper 8 (Fig. 17-4) is attracted upward and by means of link 9, connected to the ends of levers 10, moves clutch block 11 against spindle 3. Since both the clutch block and spindle have fine-pitched teeth cut on the mating surfaces, they are now solidly engaged, and continued movement of the clapper lifts the spindle against the force of the springs and releases the wheel.

(c) When the coil is deenergized, the clapper assembly drops to the disengaged position and spring pressure is applied to the shoes. As the lining wears, the spindle follows down, but the air gap between the clapper and the magnet stays the same.

(d) The disengaged position of the clapper assembly is fixed by the position of adjusting plug 12 (Fig. 17-4), which is threaded through the bottom of the housing which surrounds the clapper assembly. Unscrewing the adjusting plug will increase the travel of the spindle, and screwing it up will decrease the travel. Once this adjustment is made to obtain the proper shoe clearance, the adjusting plug should be locked by means of locknut 13.

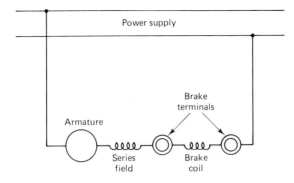

Figure 17-5   Electric motor brake with series coil. (Courtesy Westinghouse.)

Coil Connection

Refer to Figs. 17-5, 17-6, and 17-7 and note the following:

Shunt brakes are usually supplied with low-voltage coils for speedy action, and it is necessary to have a resistance in series with the coil. The power supply should be connected to BR6 for continuous rating and to BR5 for 1-hour intermittent rating. See the nameplate mounted on the brake for the correct rating.

Troubleshooting Electric Brakes

Problems attributed to the brake are often found to originate in the control. Measure the voltage at the coil terminals. On brakes using a switch or time delay for forcing the coil, check the switch or relay contacts.

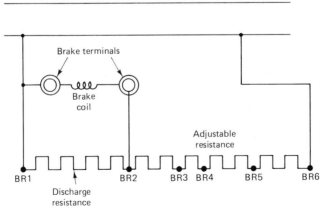

Figure 17-6   Brake with shunt coil. (Courtesy Westinghouse.)

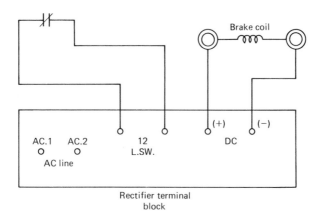

Figure 17-7 Type SA Rectox-operated motor brake. (Courtesy Westinghouse.)

Measure the spindle travel and compare with the travel specified on the nameplate. If the travel is insufficient, the shoes will not clear the wheel completely and will drag. With excessive travel, brake operation will be sluggish and shoe clearance will be greater than that required for optimum operation.

The coil may be defective because of an open condition, broken leads, or terminals. In a shorted or grounded coil, only a small part of the winding may be shorted. Check the resistance against data supplied by the manufacturer.

If the control circuits have been checked and are known to be working properly, it is safe to assume that the trouble is in the clutch assembly, provided that no broken parts can be observed on the exposed linkage and readjustment routine fails to ease the situation. Remove the magnet and clutch assembly for inspection. Look for binding parts, worn or broken bearings, loose hardware, and so on. All pivot pins running in needle bearings are hardened to Rockwell C-60. Do not attempt to make an emergency repair using a soft pin in a needle bearing.

When replacing arms or levers with new parts, it is best to order and install them in pairs. Even though these parts are machined in precision fixtures, there is still the possibility of a slight misalignment. These items are machined, stocked, and sold only in matched pairs.

After considerable service, it is possible that the serrated faces of the clutch block and rack may show some wear. If this wear reaches the point where slipping occurs, wear accelerates. For best results, both the clutch block and the rack should be replaced at the same time.

Type DI Shoe Brakes

Type DI shoe brakes are used with a dc power supply and supplied with either shunt or series coils. Type DI Rectox brakes are operated from ac power where the torque rating and/or operating characteristics of a dc brake are desired and only ac power is available. This brake is a standard type DI brake with the coil especially designed to meet the economical requirements of the rectifier unit.

The type DK brake is the type AK with a dc coil. The brake has a shunt-wound coil and is supplied in three frame sizes. The type DK brake has the same dimensions as the type AK brake for a given frame size.

Type AK shoe brakes are used with an ac power supply and are supplied with shunt-wound single-phase coils. The type AK brake features a single adjustment which restores magnet travel and compensates for wheel lining wear. Shoe clearance is automatically equalized. For higher-torque ratings either the thrustor or Rectox-operated brake is applied, depending on the operating characteristics desired.

Type AI shoe brakes are used with an ac power supply and are supplied with shunt-wound single-phase coils.

As shown in Table 17-1, the torque rating of AI solenoid brakes is limited. This is due to the excessive maintenance that would be entailed if larger torque ratings were available. For higher torque ratings either the thrustor or Rectox-operated brake is applied, depending on the operating characteristics desired.

Type HI thrustor brakes are applicable to ac power supplied where higher torque ratings are required with cushioned setting and releasing and/or adjustment time-delay setting. Thrustor brakes are operated with three-phase ac power supplied to the thrustor motor.

Disk-type brakes are available in both ac and dc power supply and are supplied only with shunt coils. Disk brakes are listed and supplied for motor mounting, as this is the most common requirement. A special motor bracket is used for motor mounting.

TABLE 17-1   Characteristics of AI and AK Brakes[a]

Frame Number	Retarding Torque (lb-ft)		Braking Capacity (hp-sec/min)	Maximum Safe Speed (rpm)
	Continuous	Intermittent		
Type AI—Class 15–340				
31	3	3	2.5	10,000
32	10	15	2.5	10,000
431	25	35	8	8,000
631	50	75	14	6,000
831	125	160	30	3,800
Type AK Brake				
41	10	15	10	10,000
43	25	35	20	8,000
73	50	75	30	5,000
103	125	160	52	3,800

[a]AI brakes are now obsolete and have been replaced by the AK line of brakes.
Source: Courtesy Westinghouse.

CHARACTERISTICS OF MOTOR BRAKE PARTS

Releasing force on SA, DI, DK, AK, AI, and disk brakes is a coil supplying sufficient pull to overcome the present spring tension. The amount of spring tension (brake torque) that the coil force can overcome is limited by the heating of the coil. The amount of force that a coil can produce is a direct function of the ampere-turns in the coil. The intermittent torque rating of a brake is considerably higher than the continuous torque rating, because more current is passed through the coil for a shorter time. Since the coil heating increases with current, only nameplate rated voltage should be applied to the coil.

The releasing force on HI brakes is a thrustor, which is a device for obtaining straight-line motion by means of oil pressure generated under a piston by a motor-driven impeller. A thrustor produces the same force whether the brake is rated as continuous or intermittent. A different torque rating is provided by changing the lever ratio.

Type SA shunt brakes are designed for 1-hour or 13-hour operation as established by the AISE Standardization Committee. These brakes are supplied with class B insulation and are designed to release at 80% of full-line voltage and to set when the voltage drops to approximately 20% of full-line voltage with coils at standard operating temperatures. A low-voltage coil and a series resistor are used to provide fast operation on shunt-wound brakes, as shown in Fig. 17-8. The discharge resistor is included in the resistor frame.

Type SA series brakes will release on 40% of full-load motor current and remain released on 10% of full-load motor current with the coil at operating temperatures. This condition applies to torque ratings for 1-hour or $\frac{1}{2}$-hour duty, which corresponds with the motor ratings. When series-wound brakes are applied to continuous-duty motors and so rated, the brakes will release at 80% of full-load motor currents and remain released on 20% or less of full-load motor current.

Type DI shunt brakes are designed to pull in at 80% of line voltage and to drop out at 20% of line voltage when the brake is properly adjusted. Usually, only a small portion of the line voltage is impressed on the coil. This is necessary to obtain snappy action of the magnet plunger, as a coil wound for full voltage results in sluggish releasing and setting of the brake.

Type DI series brakes are designed to pull in at 40% of full-load motor current and drop out at 10% of full-load motor current when properly adjusted. They are applied to carry the full-load current of the motor for their nameplate time rating.

Brakes with series coils are applied with series motors wherever possible for dc application because of the quick action and the safety factor obtained by always having sufficient current on the motor to hold the load when the brake is released. Quick action is inherent with series brakes, as the current buildup time is not greatly impeded by the coil inductance. Also, the setting time is fast because of quick current decay. Antifreeze washers are also used to further speed up release time, as in the case of the shunt brakes.

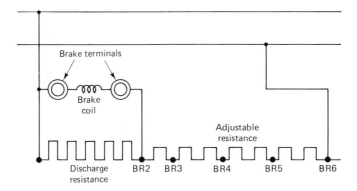

Figure 17-8 Schematic diagram of shunt brake connections SA and DI. (Courtesy Westinghouse.)

Type DI Rectox-operated brake coils are designed to operate at a minimum of 85% of ac line voltage. The coil is connected as shown in Fig. 17-9.

The limit switch mounted on top of the magnet inserts resistance and limits the coil and Rectox current when the brake is released. This allows an economical Rectox design.

The switch should be adjusted to open the circuit just before the plunger seals. If the switch is opened too soon, the reduced voltage will fail to seal the plunger,

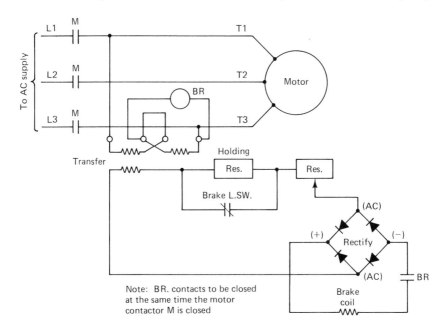

Figure 17-9 Schematic diagram for Rectox-operated brake. (Courtesy Westinghouse.)

which will then drop out and pull in with a fluttering action. To remedy this, the point of the cam that opens the switch should be trimmed to delay the opening of the switch. Care must be taken to see that when the plunger is sealed, the switch contacts are open. Failure to open will cause the Rectox unit to carry the full pull-in current, resulting in Rectox failure.

The adjustable resistor is used to compensate for rectifier aging, which results in reduced rectifier voltage output. Appreciable reduced voltage output comes only after long service, but the coil voltage should be periodically checked and kept as near as possible to its initial value by the resistor adjustment. If the voltage output becomes too low, the brake will release sluggishly or fail to release.

An auxiliary contactor BR (Fig. 17-9) is used in the brake coil circuit. This contactor opens the brake coil circuit when its coil is deenergized by the opening of the main line contactor—causing the brake to set quickly. The coil voltage is low, eliminating the necessity of a series and discharge resistor.

The dual-voltage operating transformer supplies the correct low operating voltage to the Rectox. Care should be taken when installing to make sure that the transformer is connected for the correct line voltage.

Type DS disk brake coils will operate at a minimum of 85% line voltage with an ac coil and 80% line voltage with a dc coil. All coils are wound for full voltage. All coil maintenance necessary on shoe-type brakes applies equally to disk brakes.

Types AK and A1 brake coils are designed to operate at a minimum of 85% line voltage. The shunt coils are supplied with full-line voltage as shown in Fig. 17-10.

Ac magnetic brakes are extremely fast in releasing and setting, as current buildup and decay are almost instantaneous. For this reason, no time-decreasing features such as antifreeze washers or external resistance are necessary; in addition, full-line voltage can be used.

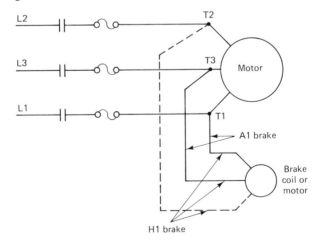

Figure 17-10 Schematic diagram for types AK, A1, or H1 brakes. (Courtesy Westinghouse.)

Because of lost impedance when an air gap is present with the plunger out of the coil, the maximum initial current that flows the instant an ac brake coil is energized is many times the amount of current flowing when the brake is fully released. More initial current will flow as the air gap increases. Therefore, it is extremely important that the plunger travel be kept close to its nameplate travel, or coil failure will result from overheating. Intermittent duty is equivalent to half time on and half time off, with continuous application of voltage not exceeding 1 hour or three operations per minute.

The clapper and plunger should be fully seated. If they are not properly seated, the air gap will cause a large current to flow, and coil failure will result from overheating.

The type HI brake thrustor is used for obtaining straight-line motion with constant pressure thrust (in one direction) from an electric motor drive.

The motors are usually the three-phase squirrel-cage type, connected as shown in Fig. 17-10, and will operate the brake satisfactorily at 85% voltage. Attention to rotation is unnecessary, because the thrustor operates correctly with either rotation. The motor bearing should be kept about one-third full of good cup grease at all times.

Oil is an important consideration, especially since the force is transmitted by it. The correct oil level must be maintained for the brake to release properly. Fill the tank until the oil starts to flow out of the oil-lever hole. When initially filling the tank, make sure that air is not trapped below the piston or in the pump by moving the piston up and down several times. If air is trapped, it will be necessary to add more oil.

With this type of brake, consideration must be given to the ambient operating temperature, because oil thickens at low temperatures. Thick oil causes overheating of the motor and sluggish operation.

Push rod seals are subject to wear. These leather seals act as oil wipers on the push rods and prevent dirt from entering the oil tank. A number of thin spacers under the retaining plate allow for tightening the packing. Tightening is done by removing from under the retaining plate one spacer at a time on each screw. The spacer should then be stacked on top of the plate under the lockwasher for further use. If repacking is necessary, remove the retaining plate and install one or more new packings on top of the old. Replace the retaining plate and the necessary number of spacers under this plate to obtain sufficient pressure for an oil-tight seal.

Care should be taken not to overtighten the seals because too much tightening would result in sluggish reclosing of the brake and rapid destruction of the brake shoe lining. This trouble should not be confused with binding at the seals caused by brake misalignment, which produces the same effect.

A time-delay feature in setting is usually supplied on thrustors and can be adjusted to provide a delay up to 10 seconds. Changing the setting time does not affect the releasing time. On some thrustors the setting time is adjusted through the filling plug. The thrustor must be energized, and while in the up position, the special wrench shipped with the thrustor is used to loosen the locknut. The setting can then be adjusted with a screwdriver. Turning the screw to the right decreases the time and to the left increases the time. On thrustors having the adjustment screw and nut on the outside of the tank cover, arrows next to the screw and nut indicate the direction of rotation for the desired time.

Thrustor brakes can be mounted at or rotated through an angle, but the thrustor must not be tilted more than 45° either side of vertical. A greater angle will result in oil leakage around the push rods and/or insufficient oil around the piston to produce proper releasing and setting.

Maintenance personnel who work on certain types of motor brakes should obtain specifications and maintenance hints from the manufacturers. Manufacturers have spent years developing good maintenance programs for their products, and such information is unavailable from any other source. Very often this information is available free of charge to users of their products, but sometimes there may be a small charge for it. The time and money spent to obtain this valuable information are well worth it. Manufacturers' names and addresses are listed in Appendix B.

appendix a

Glossary

Ac (alternating current): (1) A periodic current, the average of which is zero over a period; normally, the current reverses after given time intervals and has alternately positive and negative values; (2) the type of electrical current actually produced in a rotating generator (alternator).

Al-Cu: An abbreviation for the combination of aluminum and copper, commonly marked on terminals, lugs, and other electrical connectors to indicate that the device is suitable for use with either aluminum conductors or copper conductors.

Alloy: A composition of two or more elements, one of which is a metal and the other consisting of one or more chemical elements.

Alternator: A device to produce alternating current.

Ambient temperature: (See Temperature, ambient.)

Ammeter: An instrument used to measure current. It is calibrated in amperes.

Ampacity: The current-carrying capacity of conductors or equipment expressed in amperes.

Ampere (A): The basic SI unit measuring the quantity of electricity: that constant current that would produce between two conductors a force equal to 2×10^{-7} newton per meter of length when the current is maintained in two straight parallel conductors of infinite length of negligible cross section and placed 1 meter apart in vacuum.

Ampere-turn: The product of amperes times the number of turns in a coil.

Arc: A flow of current across an insulating medium.

Armature: (1) rotating machine: the member in which alternating voltage is generated; (2) electromagnet: the member that is moved by magnetic force.

Automatic: A device operated by its own mechanism when actuated by some impersonal influence: nonmanual: self-acting.

Automatic transfer equipment: A device to transfer a load from one power source to another, usually from normal to emergency source and back.

Auxiliary: A device or piece of equipment that aids the main device or piece of equipment.

AWG (American Wire Gage): The standard for measuring the diameter of wires in the United States; analogous to square millimeters in the SI system.

Battery: A device that converts chemical energy to electrical energy and used to store electricity.

Bond: A mechanical connection between metallic parts of an electrical system, such as between a neutral wire and a meter enclosure or service conduit to the enclosure for the service equipment with a bonding locknut or bushing; the junction of welded parts; the adhesive for abrasive grains in grinding wheels.

Bonding bushing: A special conduit bushing equipped with a conductor terminal to take a bonding jumper; also has a screw or other sharp device to bite into the enclosure wall to bond the conduit to the enclosure without a jumper when there are no concentric knockouts left in the wall of the enclosure.

Bonding jumper: A bare or insulated conductor used to provide equipment grounding continuity between conduit and cabinets or between any non-current-carrying metal enclosures of wires or electrical devices; commonly used from a bonding bushing to the service equipment enclosure to provide a path around concentric knockouts in an enclosure wall; also used to bond one raceway to another.

Bonding locknut: A threaded locknut for use on the end of a conduit terminal, but a locknut equipped with a screw through its lip. When the locknut is installed, the screw is tightened so that its end bites into the wall of the enclosure close to the edge of the knockout.

Bus: The conductor(s) serving as a common connection for two or more circuits.

Busbars: The conductive bars used as the main current supplying elements of panelboards or switchboards; also the conductive bars of copper or aluminum used within the enclosure of a busway.

Bypass: Passage at one side of or around a regular passage.

Cable, control: Used to supply voltage (usually "on" or "off").

Circuit: A closed path through which current flows from a generator, through various components, and back to the generator.

Circuit breaker: A resettable fuselike device designed to protect a circuit against overloading.

Code: Short for *National Electrical Code®*.

Coil: A wire or cable wound in a series of closed loops.

Commutator: A device used on electric motors or generators to maintain a unidirectional current.

Compressor: The pump of a refrigerating mechanism that draws a vacuum or low pressure on the cooling side of a refrigerant cycle and squeezes or compresses the gas into the high-pressure or condensing side of the cycle.

Conductance: The ability of material to carry an electric current.

Conductor: Any substance that allows energy flow through it, with the transfer being made by physical contact but excluding net mass flow.

Connection: That part of a circuit which has negligible impedance and which joins components or devices.

Contact: A device designed for repetitive connections.

Contactor: A type of relay.

Continuity: The state of being whole; unbroken.

Control: An automatic or manual device used to stop, start, and/or regulate the flow of gas, liquid, and/or electricity.

Control, temperature: A thermostatic device that automatically stops and starts a motor, the operation of which is based on temperature changes.

Controller: A device or group of devices that serves to govern in some predetermined manner the electric power delivered to the apparatus to which it is connected.

Core: The portion of a foundry mold that shapes the interior of a hollow casting.

Counter EMF: The voltage opposing the applied voltage and the current in a coil; caused by a flow of current in the coil; also known as back EMF.

Coupling: The means by which signals are transferred from one circuit to another.

Current limiting: A characteristic of short-circuit protective devices, such as fuses, by which the device operates so fast on high short-circuit currents that less than a quarter-wave of the alternating cycle is permitted to flow before the circuit is opened, thereby limiting the thermal and magnetic energy to a certain maximum value, regardless of the current available.

Cycle: (1) An interval of space or time in which one set of events or phenomena is completed; (2) a set of operations that are repeated regularly in the same sequence; (3) when a system in a given state goes through a number of different processes and finally returns to its initial state.

Direct current (dc): (1) Electricity that flows in only one direction; (2) the type of electricity produced by a battery.

Disconnect: A switch for disconnecting an electrical circuit or load (motor, transformer, panel) from the conductors that supply power to it (e.g., "He pulled the motor disconnect," meaning that he opened the disconnect switch to the motor).

Disconnecting means: A device, a group of devices, or other means whereby the conductors of a circuit can be disconnected from their supply source.

Drawing, block diagram: A simplified drawing of a system showing major items as blocks; normally used to show how the system works and the relationship between major items.

Drawing, line schematic (diagram): Shows how a circuit works.

Drawing, wiring diagram: Shows how the devices are interconnected.

Eddy currents: Circulating currents induced in conducting materials by varying magnetic fields; usually considered undesirable because they represent loss of energy and cause heating.

Efficiency: The ratio of the output to the input.

Electricity: Relating to the flow or presence of charged particles; a fundamental physical force or energy.

Electric water valve: Solenoid-type (electrically operated) valve used to turn water flow on and off.

Electrode: A conductor through which current transfers to another material.

Electromagnet: A device consisting of a ferromagnetic core and a coil that produces appreciable magnetic effects only when an electric current exists in the coil.

Electromotive force (EMF) voltage: The electrical force that causes current (free electrons) to flow or move in an electrical circuit. The unit of measurement is the volt.

Enclosed: Surrounded by a case that will prevent anyone from accidentally touching live parts.

Explosion proof: Designed and constructed to withstand an internal explosion without creating an external explosion or fire.

Fail-safe control: A device that opens a circuit when the sensing element fails to operate.

Fan: A radial or axial flow device used for moving or producing artificial currents of air.

Fault, arcing: A fault having high impedance, causing arcing.

Fault, bolting: A fault of very low impedance.

Fault, ground: A fault to ground.

Feedback: The process of transferring energy from the output circuit, of a device back to its input.

Feeder: A circuit, such as conductors in conduit or a busway run, which carries a large block of power from the service equipment to a subfeeder panel or a branch circuit panel or to some point at which the block or power is broken down into smaller circuits.

Field: The effect produced in surrounding space by an electrically charged object, by electrons in motion, or by a magnet.

Field, electrostatic: The region near a charged object.

Fitting: An accessory such as a locknut, bushing, or other part of a wiring system that is intended primarily to perform a mechanical rather than an electrical function.

Float valve: A type of valve that is operated by a sphere or pan which floats on a liquid surface and controls the level of liquid.

Fpm: Feed per minute.

Frequency: The number of complete cycles an alternating electric current, sound wave, or vibrating object undergoes per second.

Fuse: A protecting device that opens a circuit when the fusible element is severed by heating, due to overcurrent passing through. Rating: voltage, normal current, maximum let-through current, time delay of interruption.

Fuse, dual element: A fuse having two fuse characteristics; the usual combination is having an overcurrent limit and a time delay before activation.

Fuse, nonrenewable or one-time: A fuse that must be replaced after it interrupts a circuit.

Fuse, renewable link: A fuse that may be reused after current interruption by replacing the meltable link.

Fusible plug: A plug or fitting made with a metal of a known low melting temperature; used as a safety device to release pressures in case of fire.

Generator: (1) A rotating machine to convert from mechanical to electrical energy; (2) automotive–mechanical to direct current; (3) general—apparatus, equipment, and the like, to convert or change energy from one form to another.

Grommet: A plastic metal or rubber doughnut-shaped protector for wires or tubing as they pass through a hole in an object.

Ground: A large conducting body (as the earth) used as a common return for an electric circuit and as an arbitrary zero of potential.

Ground check: A pilot wire in portable cables to monitor the grounding circuit.

Ground-Fault Interrupter (GFI): A protective device that detects abnormal current flowing to ground and then interrupts the circuit.

Grounding: The device or conductor connected to ground designed to conduct only in abnormal conditions.

Grounding conductor: A conductor used to connect metal equipment enclosures and/or the system grounded conductor to a grounding electrode, such as the ground wire run to the water pipe at a service; also, may be a bare or insulated conductor used to ground motor frames, panel boxes, and other metal equipment enclosures used throughout an electrical system. In most conduit systems, the conduit is used as the grounding conductor.

Grounds: Narrow strips of wood nailed to walls as guides to plastering and as a nailing base for interior trim.

Handy box: The commonly used single-gang outlet box used for surface mounting to enclose wall switches or receptacles on concrete or cinder block construction of industrial and commercial buildings, nongangable; also made for recessed mounting; also known as "utility boxes."

Hazardous: Ignitable vapors, dust, or fibers that may cause fire or explosion.

Hertz (Hz): The derived SI unit for frequency: 1 hertz = 1 cycle per second: $1 \text{ Hz} = 1 \text{ s}^{-1}$.

Home run: That part of a branch circuit from the panelboard housing the branch circuit fuse or CB and the first junction box at which the branch circuit is spliced to lighting or receptacle devices or to conductors that continue the branch circuit to the next outlet or junction box. The term "home run" is usually reserved to multioutlet lighting and appliance circuits.

Horsepower (hp): The non-SI unit for power: 1 horsepower = 746 watts (electric) = 9800 watts (boiler).

Hot leg: A circuit conductor which normally operates at a voltage above ground; the phase wires or energized circuit wires other than a grounded neutral wire or grounded phase leg.

Hot wire: A resistance wire in an electrical relay that expands when heated and contracts when cooled.

Impedance (Z): The opposition to current flow in an ac circuit; impedance includes resistance (R), capacitive reactance (X_c), and inductive reactance (X_L); unit—ohm.

Inching: Momentary activation of machinery used for inspection or maintenance.

Induction machine: An asynchronous ac machine to change phase or frequency by converting energy—from electrical to mechanical, then from mechanical to electrical.

In phase: The condition existing when waves pass through their maximum and minimum values of like polarity at the same instant.

Instrument: A device for measuring the value of the quantity under observation.

Insulation, electrical: A medium in which it is possible to maintain an electrical field with little supply of energy from additional sources; the energy required to produce the electric field is fully recoverable only in a complete vacuum (the ideal dielectric) when the field or applied voltage is removed: used to (a) save space, (b) enhance safety, or (c) improve appearance.

Integrated circuit: A circuit in which different types of devices, such as resistors, capacitors, and transistors, are made from a single piece of material and then connected to form a circuit.

Inverter: A device that changes dc to ac.

IR (insulation resistance): The measurement of the dc resistance of insulating material; can be either volume or surface resistivity; extremely temperature sensitive.

IR drop: The voltage drop across a resistance due to the flow of current through the resistor.

Jumper: A short length of conductor, usually a temporary connection.

Junction: A connection of two or more conductors.

Junction box: A group of electrical terminals housed in a protective box or container.

Kilowatt (kw): Unit of electrical power equal to 1000 watts.

KVA: Kilovolts times amperes.

LA: See lightning arrestor.

Laminated core: An assembly of steel sheets for use as an element of magnetic circuits; the assembly has the property of reducing eddy-current losses.

Law of magnetism: Like poles repel; unlike poles attract.

Leakage: Undesirable conduction of current.

Leg: A portion of a circuit.

Lightning arrestor (LA): A device designed to protect circuits and apparatus from high transient voltage by diverting the overvoltage to ground.

Limit control: Control used to open or close electrical circuits as temperature or pressure limits are reached.

Limiter: A device in which some characteristic of the output is automatically prevented from exceeding a predetermined value.

Line: A circuit between two points: ropes used during overhead construction.

Load center: An assembly of circuit breakers or switches.

Location, damp: A location subject to a moderate amount of moisture, such as some basements, barns, cold-storage warehouses, and the like.

Location, dry: A location not normally subject to dampness or wetness; a location classified as dry may be temporarily subject to dampness or wetness, as in the case of a building under construction.

Location, wet: A location subject to saturation with water or other liquids.

Locked rotor: When the circuits of a motor are energized but the rotor is not turning.

Lockout: To keep a circuit locked open.

Lug: A device for terminating a conductor to facilitate the mechanical connection.

Magnet: A body that produces a magnetic field external to itself; magnets attract iron particles.

Magnetic field: (1) A magnetic field is said to exist at a point if a force over and above any electrostatic force is exerted on a moving charge at the point; (2) the force field established by ac through a conductor, especially a coiled conductor.

Magnetic pole: Those portions of the magnet toward which the external magnetic induction appears to converge (south) or diverge (north).

Manual: Operated by mechanical force applied directly by personal intervention.

Motor: Apparatus used to convert from electrical to mechanical energy.

Motor, capacitor: A single-phase induction motor with an auxiliary starting winding connected in series with a condenser for better starting characteristics.

Motor control: A device to start and/or stop a motor at certain temperature or pressure conditions.

Motor control center: A grouping of motor controls such as starters.

National Electrical Code® (NEC): A national consensus standard for the installation of electrical systems.

NEMA: National Electrical Manufacturers' Association.

Neutral: The element of a circuit from which other voltages are referenced with respect to magnitude and time displacement in steady-state conditions.

Neutral block: The neutral terminal block in a panelboard, meter enclosure, gutter, or other enclosure in which circuit conductors are terminated or subdivided.

Neutral wire: A circuit conductor which is common to the other conductors of the circuit, having the same voltage between it and each of the other circuit wires and usually operating grounded; such as the neutral of three-wire single-phase or three-phase four-wire wye-connected systems.

NFPA (National Fire Protection Association): An organization to promote the science and improve the methods of fire protection, which sponsors various codes, including the *National Electrical Code®*.

OHM (Ω): The derived SI unit for electrical resistance or impedance; 1 ohm equals 1 volt per ampere.

$$R = \frac{E}{I}, \quad T = \frac{E}{R}, \quad E = IR$$

Ohmmeter: An instrument for measuring resistance in ohms.

Ohm's law: A mathematical relationship between voltage, current, and resistance in an electric circuit.

OL: Overload.

Open: A circuit that is energized by not allowing useful current to flow.

OSHA (Occupational Safety and Health Act): Federal Law 91-596 of 1970, charging all employers engaged in business affecting interstate commerce to be responsible for providing a safe working place: it is administered by the Department of Labor: the OSHA regulations are published in Title 29, Chapter XVII, Part 1910 of the CFR and the *Federal Register*.

Outlet: A point on the wiring system at which current is taken to supply utilization equipment.

Overcurrent protection: Deenergizing a circuit whenever the current exceeds a predetermined value; the usual devices are fuses, circuit breakers, or magnetic relays.

Overload: A load greater than the load for which the system or mechanism was intended.

Panel: A unit for one or more sections of flat material suitable for mounting electrical devices.

Panelboard: A single panel or group of panel units designed for assembly in the form on a single panel; includes buses and may come with or without switches and/or automatic overcurrent protective devices for the control of light, heat, or power circuits of individual as well as aggregate capacity. It is designed to be placed in a cabinet or cutout box that is in or against a wall or partition and is accessible only from the front.

Parallel: Connections of two or more devices between the same two terminals of a circuit.

Phase leg: One of the phase conductors (an ungrounded or ''hot'' conductor) of a polyphase electrical system.

Plugging: Braking an induction motor by reversing the phase sequence of the power to the motor.

Pole: (1) That portion of a device associated exclusively with one electrically separated conducting path of the main circuit of the device; (2) a supporting circular column.

Positive: Connected to the positive terminal of a power supply.

Power factors: Correction coefficient for ac power necessary because of changing current and voltage values.

Pressure motor control: A device that opens and closes an electrical circuit as pressures change.

Pushbutton: A switch activated by buttons.

Raintight: So constructed or protected that exposure to a beating rain will not result in the entrance of water.

Rated: Indicating the limits of operating characteristics for application under specified conditions.

Reciprocating: Action in which the motion is back and forth in a straight line.

Rectifiers: Devices used to change alternating current to unidirectional current.

Rectify: To change from ac to dc.

Relay: A device designed to change a circuit abruptly because of a specified control input.

Relay, overcurrent: A relay designed to open a circuit when current in excess of a particular setting flows through the sensor.

Remote-control circuits: The control of a circuit through relays and similar devices.

Rheostat: A variable resistor, which can be varied while energized, normally one used in a power circuit.

Rotor: The rotating part of a mechanism.

Rpm: Revolutions per minute.

Safety motor control: An electrical device used to open a circuit if the temperature, pressure, and/or the current flow exceed safe conditions.

Sealed motor compressor: A mechanical compressor consisting of a compressor and a motor, both of which are enclosed in the same sealed housing, with no external shaft or shaft seals, and with the motor operating in the refrigerant atmosphere.

Secondary: The second circuit of a device or equipment, which is not normally connected to the supply circuit.

Sequence controls: Devices that act in series or in time order.

Service: The equipment used to connect to the conductors run from the utility line, including metering, switching, and protective devices; also the electric power delivered to the premises, rated in voltage and amperes, such as a "100-ampere 480-volt service."

Service cable: The service conductors made up in the form of a cable.

Service conductors: The supply conductors that extend from the street main or transformers to the service equipment of the premises being supplied.

Service drop: Run of cables from the power company's aerial power lines to the point of connection to a customer's premises.

Service entrance: The point at which power is supplied to a building, including the equipment used for this purpose (service main switch or panel or switchboard, metering devices, overcurrent protective devices, conductors for connecting to the power company's conductors, and raceways for such conductors).

Service equipment: The necessary equipment, usually consisting of a circuit breaker or switch and fuses and their accessories, located near the point of entrance of supply conductors to a building and intended to constitute the main control and cutoff means for the supply to that building.

Service lateral: The underground service conductors between the street main, including any risers at a pole or other structure or from transformers, and the first point of connection to the service-entrance conductors in a terminal box, meter, or other enclosure with adequate space, inside or outside the building wall. Where there is no terminal box, meter, or other enclosure with adequate space, the point of connection is the entrance point of the service conductors into the building.

Shaded-pole motor: A small dc motor used for light-start loads that has no brushes or commutator.

Short circuit: An often unintended low-resistance path through which current flows around, rather than through, a component or circuit.

Shunt: A device having appreciable resistance or impedance connected in parallel across other devices or apparatus to divert some of the current: appreciable voltage exists across the shunt and appreciable current may exist in it.

Signal: A detectable physical quantity or impulse (such as a voltage, current, or magnetic field strength) by which messages or information can be transmitted.

Signal circuit: Any electrical circuit supplying energy to an appliance that gives a recognizable signal.

Single-phase motor: An electric motor that operates on single-phase alternating current.

Single-phasing: The abnormal operation of a three-phase machine when its supply is changed by accidental opening of one conductor.

Solid state: A device, circuit, or system that does not depend on physical movement of solids, liquids, gases, or plasma.

Split-phase motor: A motor with two stator windings. Winding in use while starting is disconnected by a centrifugal switch after the motor attains speed; then the motor operates on the other winding.

Squirrel-cage motor: An induction motor having the primary winding (usually the stator) connected to the power; a current is induced in the secondary cage winding (usually the rotor).

Starter: (1) An electric controller for accelerating a motor from rest to normal speed and to stop the motor; (2) a device used to start an electric discharge lamp.

Starting relay: An electrical device that connects and/or disconnects the starting winding of an electric motor.

Starting winding: Winding in an electric motor used only during the brief period when the motor is starting.

Stator: The portion of a rotating machine that includes and supports the stationary active parts.

Switch: A device for opening and closing or for changing the connection of a circuit.

Switch, ac general-use snap: A general-use snap switch suitable only for use on alternating-current circuits and for controlling the following:

- Resistive and inductive loads (including electric discharge lamps) not exceeding the ampere rating at the voltage involved
- Tungsten-filament lamp loads not exceeding the ampere rating of the switches at the rated voltage
- Motor loads not exceeding 80% of the ampere rating of the switches at the rated voltage

Switch, ac-dc general-use snap: A type of general-use snap switch suitable for use on either direct or alternating-current circuits and for controlling the following:

- Resistive loads not exceeding the ampere rating at the voltage involved
- Inductive loads not exceeding one-half the ampere rating at the voltage involved, except that switches having a marked horsepower rating are suitable for controlling motors not exceeding the horsepower rating of the switch at the voltage involved
- Tungsten-filament lamp loads not exceeding the ampere rating at 125 volts, when marked with the letter T

Switch, general-use: A swtich intended for use in general distribution and branch circuits. It is rated in amperes and is capable of interrupting its rated voltage.

Switch, general-use snap: A type of general-use switch so constructed that it can be installed in flush device boxes or on outlet covers, or otherwise used in conjunction with wiring systems recognized by the *National Electrical Code*®.

Switch, isolating: A switch intended for isolating an electrical circuit from the source of power. It has no interrupting rating and is intended to be operated only after the circuit has been opened by some other means.

Switch, knife: A switch in which the circuit is closed by a moving blade engaging contact clips.

Switch, motor-circuit: A switch, rated in horsepower, capable of interrupting the maximum operating overload current of a motor having the same horsepower rating as the switch at the rated voltage.

Synchronism: When connected ac systems, machines, or a combination operate at the same frequency and when the phase angle displacements between voltages in them are constant, or vary about a steady and stable average value.

Synchronous: Simultaneous in action and in time (in phase).

Synchronous machine: A machine in which the average speed of normal operation is exactly proportional to the frequency of the system to which it is connected.

Synchronous speed: The speed of rotation of the magnetic flux produced by linking the primary winding.

Tachometer: An instrument for measuring revolutions per minute.

Tap: A splice connection of a wire to another wire (such as a feeder conductor in an auxiliary gutter) where the smaller conductor runs a short distance (usually only a few feet but could be up to 25 feet) to supply a panelboard or motor controller or switch. Also called a ''tap-off,'' indicating that energy is being taken from one circuit or piece of equipment to supply another circuit or load; a tool that cuts or machines threads in the side of a round hole.

Temperature, ambient: The temperature of the surrounding medium, such as air around a cable.

Terminal: A device used for connecting cables.

Torquing: Applying a rotating force and measuring or limiting its value.

Transformer, power: Designed to transfer electrical power from the primary circuit to the secondary circuit(s) to (a) step up the secondary voltage at less current, or (b) step down the secondary voltage at more current, with the voltage–current product being constant for either primary or secondary.

Universal motor: A motor designed to operate on either ac or dc at about the same speed and output with either.

Valve, solenoid: A valve actuated by magnetic action by means of an electrically energized coil.

VD: Voltage drop.

Volt (V): The derived SI unit for voltage, 1 volt equals 1 watt per ampere.

Voltage: The electrical property that provides the energy for current flow; the ratio of the work done to the value of the charge moved when a charge is moved between two points against electrical forces.

Voltage, breakdown: The minimum voltage required to break down an insulation's resistance, allowing a current flow through the insulation, normally at a point.

Voltage drop: The difference in voltage between two points of a circuit.

Voltmeter: An instrument for measuring voltage.

Watertight: So constructed that water will not enter.

Watt (W): The derived SI unit for power, radiant flux: 1 watt equals 1 joule per second.

Watt-hour (Whr): The number of watts used in 1 hour.

Watt-hour meter: A meter that measures and registers the integral, with respect to time, of the active power in a circuit.

Wattmeter: An instrument for measuring the magnitude of the active power in a circuit.

Wave: A disturbance that is a function of time or space or both.

Wet locations: Exposed to weather or water spray or buried.

Winding: An assembly of coils designed to act in consort to produce a magnetic flux field or to link a flux field.

appendix b

Trade Sources

MOTORS AND MOTOR CONTROLS

• **Contractors**

Allen Bradley Co.
1201 S. Second St.
Milwaukee, WI 53204

Arrow-Hart Inc.
Division of Crouse-Hinds
103 Hawthorn St.
Hartford, CT 06101

Automatic Switch Co.
50-56 Hanover Rd.
Florham Park, NJ 07932

Crydom
1521 Grand Av.
El Segundo, CA 90245

Cutler-Hammer Inc.
4201 N. 27th St.
Milwaukee, WI 53216

Duraline
Division of J.B. Nottingham & Co.,
    Inc.
75 Hoffman Ln.
Central Islip, NY 11722

Essex Group
131 Godfrey
Logansport, IN 46947

Federal Pacific Electric Co.
150 Av. L
Newark, NJ 07101

Furnas Electric Co.
1000 McKee St.
Batavia, IL 60510

General Electric Co.
General Purpose Control Dept.
P.O. Box 913
Bloomington, IL 61701

General Electric Co.
Industrial Control Dept.
1501 Roanoke Blvd.
Salem, VA 24153

Gould, Inc.
I-T-E Electrical Products
Rollings Meadows, IL 60008

GTE Sylvania, Inc.
Electrical Equipment Group
One Stamford Forum
Stamford, CT 06904

H-B Instrument Co.
4314 N. American St.
Philadelphia, PA 19140

Klockner-Moeller Corp.
Motor Controls
4 Strathmore Rd.
Natick, MA 01760

Mack Electric Devices Inc.
211 Glenside Av.
Wyncote, PA 19095

Magnecraft Electric Co.
5575 N. Lynch
Chicago, IL 60630

Payne Engineering Co.
Box 70
Scott Depot, WV 25560

Ross Engineering Corp.
559 Westchester Dr.
Campbell, CA 95008

Siemens-Allis
Industrial Controls Division
P.O. Box 89
Wichita Falls, TX 76307

Square D Company
P.O. Box 472
Milwaukee, WI 53201

Struthers-Dunn, Inc.
Lambs Rd.
Pitman, NJ 08071

Sylvania Electrical Control
Electrical Equipment Group
One Stamford Forum
Stamford, CT 06904

Telemecanique Inc.
2625 S. Clearbrook Dr.
Arlington Heights, IL 60005

Vectrol Inc.
110 Douglas Rd.
Oldsmar, FL 33557

Ward Leonard Electric Co., Inc.
The Unimax Group
31 South St.
Mt. Vernon, NY 10550

Westinghouse
Control Products Division
Tuscarawas Rd.
Beaver, PA 15009

Zenith Controls, Inc.
830 W. 40th St.
Chicago, IL 60609

TESTING AND MEASURING DEVICES (INSTRUMENTS)

• Ammeters

AEMC Corp.
No. Amer. Dist. Chauvin Arnoux
   Prod.
729 Boylston St.
Boston, MA 02116

Amprobe Instrument
Division of Core Industries Inc.
630 Merrick Rd.
Lynbrook NY 11563

(F.W.)Bell, Inc.
Arnold Engineering Co.
4949 Freeway Dr. E
Columbus, OH 43229

Columbia Electric Mfg. Co.
4519 Hamilton Av.
Cleveland, OH 44114

Control Power Systems Inc.
18978 NE 4th Ct.
North Miami Beach, FL 33179

Etcon Corp.
12243 S. 71st Av.
Palos Heights, IL 60463

General Electric Co.
Instrument Products Operation
40 Federal St.
Lynn, MA 01910

General Electric Co.
Instrument Rental Program
1 River Rd.
Schenectady, NY 12345

Hickok Electrical Instruments
10541 Dupont Av.
Cleveland, OH 44108

Hioki New York Corp.
46-16 235th St.
Douglaston, NY 11363

Martindale Electric Co.
1375 Hird Av.
Cleveland, OH 44107

Pacer Industries Inc.
704 E. Grand Av.
Chippewa Falls, WI 54729

RFL Industries Inc.
Powerville Rd.
Boonton, NJ 07005

Sangamo Weston, Inc.
Schlumberger Division
P.O. Box 3347
Springfield, IL 62714

Simpson Electric Co.
Division of American Gage and
    Machine Co.
853 Dundee Av.
Elgin, IL 60120

Snap-On Tools Corp.
2801 80th St.
Kenosha, WI 53140

(A.W.) Sperry Instruments Inc.
245 Marcus Bl.
Hauppauge, NY 11787

Square D Company
P.O. Box 6440
Clearwater, FL 33518

(H.H.) Sticht Co., Inc.
27 Park Pl.
New York, NY 10007

TIF Instruments Inc.
3661 NW 74th St.
Miami, FL 33147

Triplett Corp.
Bluffton, OH 45817

Western Electro Mechanical
300 Broadway
Oakland, CA 94607

Westinghouse
Relay-Instrument Division
95 Orange St., P.O. Box 606
Newark, NJ 07101

Weston Instruments
A Division of Sangamo Weston Inc.
614 Frelinghuysen Av.
Newark, NJ 07114

Yokogawa Corp. of America
5 Westchester Plaza
Elmsford, NY 10523

• Cable Tracers

Amprobe Instrument
Division of Core Industries Inc.
630 Merrick Rd.
Lynbrook, NY 11563

Aqua-Tronics Inc.
17040 SW Shaw St.
Beaverton, OR 97005

Associated Research Inc.
8221 N. Kimball Av.
Skokie, IL 60076

(James G.) Biddle Co.
Township Line & Jolly Rds.
Plymouth Meeting, PA 19462

Cranleigh Electro-Thermal Inc.
P.O. Box 7500
Menlo Park, CA 94025

FRL Inc.
Fisher Div.
517 Marine View Av.
Belmont, CA 94002

General Electric Co.
Instrument Rental Program
1 River Rd.
Schenectady, NY 12345

Goldak Co., Inc.
626 Sonora Av.
Glendale, CA 91201

Hipotronics
Route 22
Brewster, NY 10509

Metrotech Corp.
670 National Av.
Mountain View, CA 94043

Radar Engineers
Division of Epic Corp.
4654 NE Columbia Blvd.
Portland, OR 97218

Rycom Instruments Inc.
9351 E. 59th St.
Raytown, MO 64133

Systems Research, Inc.
P.O. Box 25280
Portland, OR 97225

TIF Instruments Inc.
3661 NW 74th St.
Miami, FL 33147

Tinker & Rasor
P.O. Box 281
San Gabriel, CA 91778

Utility Products
Division of Maxwell Laboratories
   Inc.
8835 Balboa Av.
San Diego, CA 92123

Western Progress
835 Maude Av.
Mountain View, CA 94043

• Circuit Breaker Testers
Anderson Power Products Inc.
145 Newton St.
Boston, MA 02135

(James G.) Biddle Co.
Township Line & Jolly Rds.
Plymouth Meeting, PA 19462

Electroware Products Inc.
24 Lisa Dr.
Dix Hills, NY 11746

General Electric Co.
Instrument Rental Program
1 River Rd.
Schenectady, NY 12345

Hioki New York Corp.
46-16 235th St.
Douglaston, NY 11363

Hipotronics, Inc.
Route 22
Brewster, NY 10509

Leviton Mfg. Co., Inc.
59-25 Little Neck Pwy.
Little Neck, NY 11362

TIF Instruments Inc.
3661 NW 74th St.
Miami, FL 33147

• Continuity Testers
AEMC Corp.
No. Amer. Dist. Chauvin Arnoux
   Prod.
729 Boylston St.
Boston, MA 02116

Amprobe Instrument
Division of Core Industries Inc.
630 Merrick Rd.
Lynbrook, NY 11563

Associated Research Inc.
8221 N. Kimball Av.
Skokie, IL 60076

Beckman Research & Mfg. Corp.
111 W. Ash Av.
Burbank, CA 91502

(James G.) Biddle Co.
Township Line & Jolly Rds.
Plymouth Meeting, PA 19462

(W.H.) Brady Co.
Division of Industrial Products
2221 W. Camden Rd.
Milwaukee, WI 53201

Bright Star Industries Inc.
600 Getty Av.
Clifton, NJ 07015

Burnworth Tester Co.
815 Ponoma Av.
El Cerrito, CA 94530

Control Power Systems Inc.
18978 NE 4th Ct.
North Miami Beach, FL 33179

Ecos Electronics Corp.
205 W. Harrison St.
Oak Park, IL 60304

Electroware Products Inc.
24 Lisa Dr.
Dix Hills, NY 11746

Etcon Corp.
12243 S. 71st Av.
Palos Heights, IL 60463

Ft. Wayne Electrical Center
Division of Weingart Inc.
1800 Broadway
Ft. Wayne, IN 46804

General Electric Co.
Instrument Products Operation
40 Federal St.
Lynn, MA 01910

Genisco Tech. Corp.
18435 Susana Rd.
Compton, CA 90221

Hipotronics Inc.
Route 22
Brewster, NY 10509

Ideal Industries, Inc.
5224 Becker Pl.
Sycamore, IL 60178

ITT Holub Industries
443 Elm St.
Sycamore, IL 60178

Martindale Electric Co.
1375 Hird Av.
Cleveland, OH 44107

Paragon Electric Co. Inc.
AMF Incorporated
606 Parkway Blvd.
Box 28
Two Rivers, WI 54241

Peschel Instruments Inc.
1412 Viscaya Pkwy.
Cape Coral, FL 33904

Ric-Nor Co. Inc.
10 Roland St.
Charlestown, MA 02129

Simpson Electric Co.
Division of American Gage and
  Machine Co.
853 Dundee Av.
Elgin, IL 60120

Snap-On Tools Corp.
2801 80th St.
Kenosha, WI 53140

(A.W.) Sperry Instruments, Inc.
245 Marcus Bl.
Hauppauge, NY 11787

(H.H.) Sticht Co., Inc.
27 Park Pl.
New York, NY 10007

Systems Research, Inc.
P.O. Box 25280
Portland, OR 97225

Teledyne Penn-Union Monarch Fuse
Division of Teledyne Inc.
229 Waterford St.
Edinboro, PA 16412

Techni-Tool Inc.
Apollo Rd.
Plymouth Meeting, PA 19462

TIF Instruments, Inc.
3661 NW 74th St.
Miami, FL 33147

Triplett Corp.
Bluffton, OH 45817

Vaco Products Co.
1510 Skokie Blvd.
Northbrook, IL 60062

Woodhead Co., Daniel
Division of Daniel Woodhead Inc.
3411 Woodhead Dr.
Northbrook, IL 60062

• Dynamometers

(W.C.) Dillon & Co., Inc.
14620 Keswick St.
Van Nuys, CA 91407

General Electric Co.
Direct Current Motor & Generator
  Dept.
3001 E. Lake Rd.
Erie, PA 16531

Linemen's Supply
Division of Buckingham Mfg. Co.,
  Inc.
7-9 Travis Av.
Binghamton, NY 13904

Simpson Electric Co.
Division of American Gage and
  Machine Co.
853 Dundee Av.
Elgin, IL 60120

Techni-Tool Inc.
Apollo Rd.
Plymouth Meeting, PA 19462

Weston Instruments
A Division of Sangamo Weston Inc.
614 Frelinghuysen Av.
Newark, NJ 07114

• Fault Indicators, URD

Associated Research Inc.
8221 N. Kimball Av.
Skokie, IL 60076

Burndy Corp.
Richards Av.
Norwalk, CT 06856

(James G.) Biddle Co.
Township Line & Jolly Rds.
Plymouth Meeting, PA 19462

Ecos Electronics Corp.
205 W. Harrison St.
Oak Park, IL 60304

Genisco Technology Corp.
18435 Susana Rd.
Compton, CA 90221

Kabo Electronics
123 Bacon
Natick, MA 01760

• Fault Locators
AEMC Corp.
No. Amer. Dist. Chauvin Arnoux
    Prod.
729 Boylston St.
Boston, MA 02116

Aqua-Tronics Inc.
17040 SW Shaw St.
Beaverton, OR 97005

Associated Research Inc.
8221 N. Kimball Av.
Skokie, IL 60076

(James G.) Biddle Co.
Township Line & Jolly Rds.
Plymouth Meeting, PA 19462

Ecos Electronics Corp.
205 W. Harrison St.
Oak Park, IL 60304

Fisher Research Laboratory
Fisher Division
517 Marine View Av.
Belmont, CA 94002

Goldak Co., Inc.
626 Sonora Av.
Glendale, CA 91201

Hipotronics Inc.
Route 22
Brewster, NY 10509

Kabo Electronics
123 Bacon
Natick, MA 01760

Peschel Instruments Inc.
1412 Viscaya Pkwy.
Cape Coral, FL 33904

Progressive Electronics Inc.
432 S. Extension Rd.
Mesa, AZ 85202

Rycom Instruments Inc.
9351 E. 59th St.
Raytown, MO 64133

Sotcher Measurement Inc.
1120 J. Stewart Court
Sunnyvale, CA 94086

Square D. Company
P.O. Box 6440
Clearwater, FL 33518

Systems Research, Inc.
P.O. Box 25280
Portland, OR 97225

TIF Instruments, Inc.
3661 NW 74th St.
Miami, FL 33147

Utility Products
Division of Maxwell Laboratories,
    Inc.
8835 Balboa Av.
San Diego, CA 92123

(The) Von Corp.
P.O. Box 3566G
Birmingham, AL 35211

Western Progress
835 Maude Av.
Mountain View, CA 94043

• GFI Circuit Testers
Daltec Systems Inc.
Box 157 (Onondaga Br.)
Syracuse, NY 13215

Ecos Electronics Corp.
205 W. Harrison St.
Oak Park, IL 60304

Etcon Corporation
12243 S. 71st Av.
Palos Heights, IL 60463

Genisco Technology Corp.
18435 Susana Rd.
Compton, CA 90221

Harvey Hubbell Inc.
Wiring Device Division
State St.
Bridgeport, CT 06602

Kabo Electronics
123 Bascon
Natick, MA 01760

Leviton Mfg. Co. Inc.
59-25 Little Neck Pwy
Little Neck, NY 11362

Sotcher Measurement Inc.
1120 J Stewart Court
Sunnyvale, CA 94086

Triple S Products
200 E. Prairie St.
Vicksburg, MI 49097

Wadsworth Electric Mfg. Co. Inc.
P.O. Box 272
Covington, KY 41012

• Ground Detectors
Associated Research Inc.
8221 N. Kimball Av.
Skokie, IL 60076

Burnworth Tester Co.
815 Pomona Av.
El Cerrito, CA 94530

Eagle Electric Mfg. Co. Inc.
45-31 Court Square
Long Island City, NY 11101

Ecos Electronics Corp.
205 W. Harrison St.
Oak Park, IL 60304

Erickson Electrical Equipment Co.
4460 N. Elston Av.
Chicago, IL 60630

General Electric Co.
Instrument Rental Program
1 River Rd.
Schenectady, NY 12345

Hipotronics Inc.
Route 22
Brewster, NY 10509

Kabo Electronics
123 Bacon
Natick, MA 01760

Key Systems Inc.
Allenwood-Herbertsville Rd.
Howell, NJ 07731

Midland-Ross Corp.
Electrical Products Div.
530 W. Mt. Pleasant Av.
Livingston, NJ 07039

Rochester Instrument Systems Inc.
255 N. Union St.
Rochester, NY 14605

Russellstoll
Electrical Products Division
Midland-Ross Corp.
530 W. Mt. Pleasant Av.
Livingston, NJ 07039

Sotcher Measurement Inc.
1120 J. Stewart Court
Sunnyvale, CA 94086

Wadsworth Electric Mfg. Co. Inc.
P.O. Box 272
Covington, KY 41012

Westinghouse
Relay-Instrument Division
95 Orange St.
P.O. Box 606
Newark, NJ 07101

• Ground Testers
AEMC Corporation
No. Amer. Dist. Chauvin Arnoux
   Prod.
729 Boylston St.
Boston, MA 02116

AMP Special Industries
Division of AMP Products Corp.
Valley Forge, PA 19482

Associated Research Inc.
8221 N. Kimball Av.
Skokie, IL 60076

Burnworth Tester Co.
815 Pomona Av.
El Cerrito, CA 94530

Eagle Electric Mfg. Co., Inc.
45-31 Court Square
Long Island City, NY 11101

Ecos Electronics Corp.
205 W. Harrison St.
Oak Park, IL 60304

General Electric Co.
Instrument Rental Program
1 River Rd.
Schenectady, NY 12345

Genisco Technology Corp.
18435 Susana Rd.
Compton, CA 90221

Hioki New York Corp.
46-16 235th St.
Douglaston, NY 11363

Hipotronics, Inc.
Route 22
Brewster, NY 10509

Independent Protection Co. Inc.
1603-09 S. Main St.
Goshen, IN 46526

Ric-Nor Co., Inc.
10 Roland St.
Charlestown, MA 02129

Russellstoll
Division of Electrical Products
Midland-Ross Corp.
530 W. Mt. Pleasant Av.
Livingston, NJ 07039

Simpson Electric Co.
American Gage and Machine Co.
   Division
853 Dundee Av.
Elgin, IL 60120

Sotcher Measurement Inc.
1120 J. Stewart Court
Sunnyvale, CA 94086

(H.H.) Sticht Co., Inc.
27 Park Pl.
New York, NY 10007

TIF Instruments Inc.
3661 NW 74th St.
Miami, FL 33147

Triplett Corp.
Bluffton, OH 45817

(Daniel) Woodhead Co.
Division of Daniel Woodhead Inc.
3411 Woodhead Dr.
Northbrook, IL 60062

Yokogawa Corp. of America
5 Westchester Plaza
Elmsford, NY 10523

• **Insulation Testers**
AEMC Corp.
No. Amer. Dist. Chauvin Arnoux
   Prod.
729 Boylston St.
Boston, MA 02116

Amprobe Instrument
Division of Core Industries Inc.
630 Merrick Rd.
Lynbrook, NY 11563

Associated Research Inc.
8221 N. Kimball Av.
Skokie, IL 60076

Beckman Instruments Inc.
Division of Cedar Grove Operations
89 Commerce Rd.
Cedar Grove, NJ 07009

(James G.) Biddle Co.
Township Line & Jolly Rds.
Plymouth Meeting, PA 19462

Burnworth Tester Co.
815 Pomona Av.
El Cerrito, CA 94530

Ecos Electronics Corp.
205 W. Harrison St.
Oak Park, IL 60304

Ft. Wayne Electrical Center
Division of Weingart, Inc.
1800 Broadway
Ft. Wayne, IN 46804

Hioki New York Corp.
46-16 235th St.
Douglaston, NY 11363

Hipotronics, Inc.
Route 22
Brewster, NY 10509

Martindale Electric Co.
1375 Hird Av.
Cleveland, OH 44107

Peschel Instruments Inc.
1412 Viscaya Pkwy.
Cape Coral, FL 33904

Ricca-Reddington Instruments, Inc.
1400G NW 65th Av.
Plantation, FL 33313

Ross Engineering Corp.
559 Westchester Dr.
Campbell, CA 95008

Simpson Electric Co.
Division of American Gage and
   Machine Co.
853 Dundee Av.
Elgin, IL 60120

Sotcher Measurement Inc.
1120 J Stewart Court
Sunnyvale, CA 94086

(A.W.) Sperry Instruments Inc.
245 Marcus Bl.
Hauppauge, NY 11787

(H.H.) Sticht Co., Inc.
27 Park Pl.
New York, NY 10007

TIF Instruments, Inc.
3661 NW 74th St.
Miami, FL 33147

(The) Von Corp.
P.O. Box 3566G
Birmingham, AL 35211

(Daniel) Woodhead Co.
Division of Daniel Woodhead Inc.
3411 Woodhead Dr.
Northbrook, IL 60062

Yokogawa Corp. of America
5 Westchester Plaza
Elmsford, NY 10523

• **Line Testers**
Beckman Research & Mfg. Corp.
111 W. Ash Av.
Burbank, CA 91502

(James G.) Biddle Co.
Township Line & Jolly Rds.
Plymouth Meeting, PA 19462

(W.H.) Brady Co.
Industrial Products Division
2221 W. Camden Rd.
Milwaukee, WI 53201

Burnworth Tester Co.
815 Pomona Av.
El Cerrito, CA 94530

Ecos Electronics Corp.
205 W. Harrison St.
Oak Park, IL 60304

Etcon Corporation
12243 S. 71st Av.
Palos Heights, IL 60463

Electroware Products Inc.
24 Lisa Dr.
Dix Hills, NY 11746

Gardner Bender, Inc.
6101 N. Baker Rd.
P.O. Box 23322
Milwaukee, WI 53209

General Electric Co.
Instrument Products Operation
40 Federal St.
Lynn, MA 01910

Genisco Technology Corp.
18435 Susana Rd.
Compton, CA 90221

Hipotronics, Inc.
Route 22
Brewster, NY 10509

Pacer Industries Inc.
704 E. Grand Av.
Chippewa Falls, WI 54729

Progressive Electronics Inc.
432 S. Extension Rd.
Mesa, AZ 85202

Ric-Nor Co., Inc.
10 Roland St.
Charlestown, MA 02129

Rodale Mfg./Square D Co.
Sixth & Minor Sts.
Emmaus, PA 18049

TIF Instruments Inc.
3661 NW 74th St.
Miami, FL 33147

Time Mark Corp.
P.O. Box 15127
Tulsa, OK 74115

Triplett Corp.
Bluffton, OH 45817

Vaco Products Co.
1510 Skokie Blvd.
Northbrook, IL 60062

• **Measuring Wheels**
Bessemer Manufacturing Corp.
412 W. King St.
York, PA 17405

(James G.) Biddle Co.
Township Line & Jolly Rds.
Plymouth Meeting, PA 19462

Linemen's Supply
Division of Buckingham Mfg. Co.,
    Inc.
7-9 Travis Av.
Binghamton, NY 13904

Rolatape Corp.
4221 Redwood Av.
Los Angeles, CA 90066

• **Megohmmeters**
AEMC Corp.
No. Amer. Dist. Chauvin Arnoux
   Prod.
729 Boylston St.
Boston, MA 02116

Amprobe Instrument
Division of Core Industries Inc.
630 Merrick Rd.
Lynbrook, NY 11563

Associated Research Inc.
8221 N. Kimball Av.
Skokie, IL 60076

Beckman Instruments Inc.
Division of Cedar Grove Operations
89 Commerce Rd.
Cedar Grove, NJ 07009

Beckman Research & Mfg. Corp.
111 W. Ash Av.
Burbank, CA 91502

(James G.) Biddle Co.
Township Line & Jolly Rds.
Plymouth Meeting, PA 19462

Ecos Electronics Corp.
205 W. Harrison St.
Oak Park, IL 60304

General Electric Co.
Instrument Rental Program
1 River Rd.
Schenectady, NY 12345

Hickok Electrical Instruments
10514 Dupont Av.
Cleveland, OH 44108

Hioki New York Corp.
46-16 235th St.
Douglaston, NY 11363

Hipotronics, Inc.
Route 22
Brewster, NY 10509

Martindale Electric Co.
1375 Hird Av.
Cleveland, OH 44107

Pacer Industries Inc.
704 E. Grand Av.
Chippewa Falls, WI 54729

Peschel Instruments Inc.
1412 Viscaya Pkwy.
Cape Coral, FL 33904

Reliant Heating & Controls Inc.
3433 Edward Av.
Santa Clara, CA 95050

Ricca-Reddington Instruments, Inc.
1400G NW 65th Av.
Plantation, FL 33313

Ross Engineering Corp.
559 Westchester Dr.
Campbell, CA 95008

Simpson Electric Co.
Division of American Gage and
   Machine Co.
853 Dundee Av.
Elgin, IL 60120

(A.W.) Sperry Instruments Inc.
245 Marcus Bl.
Hauppauge, NY 11787

(H.H.) Sticht Co., Inc.
27 Park Pl.
New York, NY 10007

TIF Instruments Inc.
3661 NW 74th St.
Miami, FL 33147

Triplett Corporation
Bluffton, OH 45817

(The) Von Corp.
P.O. Box 3566G
Birmingham, AL 35211

Weston Instruments
A Division of Sangamo Weston Inc.
614 Frelinghuysen Av.
Newark, NJ 07114

(Daniel) Woodhead Co.
A Division of Daniel Woodhead Inc.
3411 Woodhead Dr.
Northbrook, IL 60062

Yokogawa Corp. of America
5 Westchester Plaza
Elmsford, NY 10523

• **Meters, Phase Angle**
Beckwith Electric Co., Inc.
11811 62nd St.
North Largo, FL 33543

Time Mark Corp.
P.O. Box 15127
Tulsa, OK 74115

• Meters, Relative Humidity
Abbeon Cal Inc.
123 Gray Av.
Santa Barbara, CA 93101

Beckman Instruments Inc.
Division of Cedar Grove Operations
89 Commerce Rd.
Cedar Grove, NJ 07009

Epic Inc.
150 Nassau St.
New York, NY 10038

H-B Instrument Co.
4314 N. American St.
Philadelphia, PA 19140

• Ohmmeters
AEMC Corporation
No. Amer. Dist. Chauvin Arnoux
  Prod.
729 Boylston St.
Boston, MA 02116

Amprobe Instrument
Division of Core Industries Inc.
630 Merrick Rd.
Lynbrook, NY 11563

Anderson Power Products Inc.
145 Newton St.
Boston, MA 02135

Associated Research Inc.
8221 N. Kimball Av.
Skokie, IL 60076

Beckman Instruments Inc.
Division of Cedar Grove Operations
89 Commerce Rd.
Cedar Grove, NJ 07009

(James G.) Biddle Co.
Township Line & Jolly Rds.
Plymouth Meeting, PA 19462

Burnworth Tester Co.
815 Pomona Av.
El Cerrito, CA 94530

Ecos Electronics Corporation
205 W. Harrison St.
Oak Park, IL 60304

Etcon Corporation
12243 S. 71st Av.
Palos Heights, IL 60463

General Electric Co.
Instrument Products Operation
40 Federal St.
Lynn, MA 01910

Hickok Electrical Instruments
10514 Dupont Av.
Cleveland, OH 44108

Martindale Electric Co.
1375 Hird Av.
Cleveland, OH 44107

Non-Linear Systems Inc.
533 Stevens Av.
Solana Beach, CA 92075

Ross Engineering Corporation
559 Westchester Dr.
Campbell, CA 95008

Sangamo Weston, Inc.
Schlumberger Division
P.O. Box 3347
Springfield, IL 62714

Snap-On Tools Corp.
2801 80th St.
Kenosha, WI 53140

(A.W.) Sperry Instruments Inc.
245 Marcus Bl.
Hauppauge, NY 11787

(H.H.) Sticht Co., Inc.
27 Park Pl.
New York, NY 10007

TIF Instruments Inc.
3661 NW 74th St.
Miami, FL 33147

Triplett Corp.
Bluffton, OH 45817

Weston Instruments
A Division of Sangamo Weston Inc.
614 Frelinghuysen Av.
Newark, NJ 07114

(Daniel) Woodhead Co.
Division of Daniel Woodhead Inc.
3411 Woodhead Dr.
Northbrook, IL 60062

Yokogawa Corp. of America
5 Westchester Plaza
Elmsford, NY 10523

• Phase Sequence Indicators

AEMC Corp.
No. Amer. Dist. Chauvin Arnoux
  Prod.
729 Boylston St.
Boston, MA 02116

Amprobe Instrument
Division of Core Industries Inc.
630 Merrick Rd.
Lynbrook, NY 11563

Applied Electro Technology
2220 S. Anne St.
Santa Ana, CA 92704

Associated Research Inc.
8221 N. Kimball Av.
Skokie, IL 60076

(James G.) Biddle Co.
Township Line & Jolly Rds.
Plymouth Meeting, PA 19462

Ecos Electronics Corp.
205 W. Harrison St.
Oak Park, IL 60304

Epic Inc.
150 Nassau St.
New York, NY 10038

General Electric Co.
Instrument Rental Program
1 River Rd.
Schenectady, NY 12345

General Equipment & Mfg. Co. Inc.
3300 Fern Valley Rd.
Louisville, KY 40213

Hipotronics
Route 22
Brewster, NY 10509

Knopp Inc.
1307 66th St.
Oakland, CA 94608

Lark Electronics Inc.
390 Ft. George Sta.
New York, NY 10040

Martindale Electric Co.
1375 Hird Av.
Cleveland, OH 44107

(H.H.) Sticht Co., Inc.
27 Park Pl.
New York, NY 10007

Time Mark Corp.
P.O. Box 15127
Tulsa, OK 74115

Western Electro Mechanical
300 Broadway
Oakland, CA 94607

Westinghouse
Division of Relay-Instrument
95 Orange St., P.O. Box 606
Newark, NJ 07101

• Power Factor Meters

AEMC Corp.
No. Amer. Dist. Chauvin Arnoux
  Prod.
729 Boylston St.
Boston, MA 02116

(James G.) Biddle Co.
Township Line & Jolly Rds.
Plymouth Meeting, PA 19462

Ecos Electronics Corp.
205 W. Harrison St.
Oak Park, IL 60304

Epic Inc.
150 Nassau St.
New York, NY 10038

General Electric Co.
Instrument Products Operation
40 Federal St.
Lynn, MA 01910

General Electric Co.
Instrument Rental Program
1 River Rd.
Schenectady, NY 12345

Martindale Electric Co.
1375 Hird Av.
Cleveland, OH 44107

Square D Company
P.O. Box 6440
Clearwater, FL 33518

Westinghouse
Relay-Instrument Division
95 Orange St.
P.O. Box 606
Newark, NJ 07101

Yokogawa Corporation of America
5 Westchester Plaza
Elmsford, NY 10523

• Recorders, Volt, Amp, Temp.

Amprobe Instrument
Division of Core Industries Inc.
630 Merrick Rd.
Lynbrook, NY 11563

Duraline
Division of J.B. Nottingham & Co.,
  Inc.
75 Hoffman Ln.
Central Islip, NY 11722

General Electric Co.
Instrument Products Operation
40 Federal St.
Lynn, MA 01910

General Electric Co.
Instrument Rental Program
1 River Rd.
Schenectady, NY 12345

Genisco Technology Corp.
18435 Susana Rd.
Compton, CA 90221

Hastings Fiber Glass Prods. Inc.
770 S. Cook Rd.
Hastings, MI 49058

Martindale Electric Co.
1375 Hird Av.
Cleveland, OH 44107

Pacer Industries Inc.
704 E. Grand Av.
Chippewa Falls, WI 54729

Photron Instrument Co.
6516 Detroit Av.
Cleveland, OH 44102

Reliant Heating & Controls Inc.
3433 Edward Av.
Santa Clara, CA 95050

Sangamo Weston Inc.
Schlumberger Division
P.O. Box 3347
Springfield, IL 62714

Simpson Electric Co.
American Gage and Machine Co.
853 Dundee Av.
Elgin, IL 60120

(H.H.) Sticht Co., Inc.
27 Park Pl.
New York, NY 10007

Westinghouse
Relay-Instrument Division
95 Orange St.
P.O. Box 606
Newark, NJ 07101

Yokogawa Corp. of America
5 Westchester Plaza
Elmsford, NY 10523

• Tachometers

Abbeon Cal Inc.
123 Gray Av.
Santa Barbara, CA 93101

AEMC Corp.
No. Amer. Dist. Chauvin Arnoux
  Prod.
729 Boylston St.
Boston, MA 02116

(James G.) Biddle Co.
Township Line & Jolly Rds.
Plymouth Meeting, PA 19462

Dynalco Corp.
5200 NW 37th Av.
Ft. Lauderdale, FL 33310

Ecos Electronics Corp.
205 W. Harrison St.
Oak Park, IL 60304

Epic Inc.
150 Nassau St.
New York, NY 10038

General Electric Co.
Instrument Products Operation
40 Federal St.
Lynn, MA 01910

Martindale Electric Co.
1375 Hird Av.
Cleveland, OH 44107

Pacer Industries Inc.
704 E. Grand Av.
Chippewa Falls, WI 54729

Simpson Electric Co.
American Gage and Machine Co.
  Division
853 Dundee Av.
Elgin, IL 60120

Snap-On Tools Corp.
2801 80th St.
Kenosha, WI 53140

(H.H.) Sticht Co., Inc.
27 Park Pl.
New York, NY 10007

TIF Instruments Inc.
3661 NW 74th St.
Miami, FL 33147

Weston Instruments
A Division of Sangamo Weston Inc.
614 Frelinghuysen Av.
Newark, NJ 07114

• **Temp. Measuring**
Amprobe Instrument
Division of Core Industries Inc.
630 Merrick Rd.
Lynbrook, NY 11563

(James G.) Biddle Co.
Township Line & Jolly Rds.
Plymouth Meeting, PA 19462

Duraline
Division of J.B. Nottingham & Co.,
   Inc.
75 Hoffman Ln.
Central Islip, NY 11722

Ecos Electronics Corp.
205 W. Harrison St.
Oak Park, IL 60304

Fenwal Inc.
Div. of Walter Kidde & Co., Inc.
400 Main St.
Ashland, MA 01721

General Electric Co.
Instrument Products Operation
40 Federal St.
Lynn, MA 01910

Genisco Technology Corp.
18435 Susana Rd.
Compton, CA 90221

(Claud S.) Gordon Co.
5710 Kenosha St.
Richmond, IL 60071

H-B Instrument Co.
4314 N. American St.
Philadelphia, PA 19140

Hy-Cal Engineering
12105 Los Nietos Rd.
Santa Fe Springs, CA 90670

ITT Holub Industries
443 Elm St.
Sycamore, IL 60178

Mack Electric Devices Inc.
211 Glenside Av.
Wyncote, PA 19095

Mikron Instrument Co., Inc.
445 W. Main St.
Wyckoff, NJ 07481

Omron Electronics Inc.
233 S. Wacker Dr., #5300
Chicago, IL 60606

Payne Engineering Co.
Box 70
Scott Depot, WV 25560

RFL Industries Inc.
Powerville Rd.
Boonton, NJ 07005

Simpson Electric Co.
Division of American Gage and
   Machine Co.
853 Dundee Av.
Elgin, IL 60120

Techni-Tool Inc.
Apollo Rd.
Plymouth Meeting, PA 19462

TIF Instruments Inc.
3661 NW 74th St.
Miami, FL 33147

Triplett Corp.
Bluffton, OH 45817

Weston Instruments
A Division of Sangamo Weston Inc.
614 Frelinghuysen Av.
Newark, NJ 07114

• **Thermometers**
Abbeon Cal Inc.
123 Gray Av.
Santa Barbara, CA 93101

AEMC Corp.
No. Amer. Dist. Chauvin Arnoux
   Prod.
729 Boylston St.
Boston, MA 02116

Amprobe Instrument
Division of Core Industries Inc.
630 Merrick Rd.
Lynbrook, NY 11563

Fenwal Inc.
Division of Walter Kidde & Co., Inc.
400 Main St.
Ashland, MA 01721

(Claud S.) Gordon Co.
5710 Kenosha St.
Richmond, IL 60071

H-B Instrument Co.
4314 N. American St.
Philadelphia, PA 19140

Hy-Cal Engineering
12105 Los Nietos Rd.
Santa Fe Springs, CA 90670

Mack Electric Devices Inc.
211 Glenside Av.
Wyncote, PA 19095

Sagline Inc.
P.O. Box 351
Millwood, NY 10546

TIF Instruments Inc.
3661 NW 74th St.
Miami, FL 33147

United Electric Controls Co.
85 School St.
Watertown, MA 02172

Weston Instruments
A Division of Sangamo Weston Inc.
614 Frelinghuysen Av.
Newark, NJ 07114

• Volt-Ammeters
AEMC Corp.
No. Amer. Dist. Chauvin Arnoux
   Prod.
729 Boylston St.
Boston, MA 02116

Amprobe Instrument
Division of Core Industries Inc.
630 Merrick Rd.
Lynbrook, NY 11563

Associated Research Inc.
8221 N. Kimball Av.
Skokie, IL 60076

Burnworth Tester Co.
815 Pomona Av.
El Cerrito, CA 94530

Columbia Electric Mfg. Co.
4519 Hamilton Av.
Cleveland, OH 44114

Control Power Systems Inc.
18978 NE 4th Ct.
North Miami Beach, FL 33179

Ecos Electronics Corp.
205 W. Harrison St.
Oak Park, IL 60304

Epic Inc.
150 Nassau St.
New York, NY 10038

Etcon Corp.
12243 S. 71st Av.
Palos Heights, IL 60463

General Electric Co.
Instrument Products Operation
40 Federal St.
Lynn, MA 01910

General Electric Co.
Instrument Rental Program
1 River Rd.
Schenectady, NY 12345

Hickok Electrical Instruments
10514 Dupont Av.
Cleveland, OH 44108

Hioki New York Corp.
46-16 235th St.
Douglaston, NY 11363

ITT Holub Industries
443 Elm St.
Sycamore, IL 60178

Pacer Industries Inc.
704 E. Grand Av.
Chippewa Falls, WI 54729

Sangamo Weston Inc.
Schlumberger Division
P.O. Box 3347
Springfield, IL 62714

Simpson Electric Co.
American Gage and Machine Co.
   Division
853 Dundee Av.
Elgin, IL 60120

Snap-On Tools Corp.
2801 80th St.
Kenosha, WI 53140

(A.W.) Sperry Instruments Inc.
245 Marcus Bl.
Hauppauge, NY 11787

(H.H.) Sticht Co., Inc.
27 Park Pl.
New York, NY 10007

TIF Instruments Inc.
3661 NW 74th St.
Miami, FL 33147

Triplett Corp.
Bluffton, OH 45817

Viz Mfg. Co.
335 E. Price St.
Philadelphia, PA 19144

Western Electro Mechanical
300 Broadway
Oakland, CA 94607

Weston Instruments
A Division of Sangamo Weston Inc.
614 Frelinghuysen Av.
Newark, NJ 07114

Yokogawa Corp. of America
5 Westchester Plaza
Elmsford, NY 10523

• **Voltmeters**
AEMC Corp.
No. Amer. Dist. Chauvin Arnoux
   Prod.
729 Boylston St.
Boston, MA 02116

Amprobe Instrument
Division of Core Industries Inc.
630 Merrick Rd.
Lynbrook, NY 11563

Associated Research Inc.
8221 N. Kimball Av.
Skokie, IL 60076

Burnworth Tester Co.
815 Pomona Av.
El Cerrito, CA 94530

Columbia Electric Mfg. Co.
4519 Hamilton Av.
Cleveland, OH 44114

Control Power Systems Inc.
18978 NE 4th Ct.
North Miami Beach, FL 33179

Ecos Electronics Corp.
205 W. Harrison St.
Oak Park, IL 60304

Etcon Corp.
12243 S. 71st Av.
Palos Heights, IL 60463

Gardner Bender Inc.
6101 N. Baker Rd.
P.O. Box 23322
Milwaukee, WI 53209

General Electric Co.
Instrument Products Operation
40 Federal St.
Lynn, MA 01910

General Electric Co.
Instrument Rental Program
1 River Rd.
Schenectady, NY 12345

Hickok Electrical Instruments
10514 Dupont Av.
Cleveland, OH 44108

Hioki New York Corp.
46-16 235th St.
Douglaston, NY 11363

Hipotronics Inc.
Route 22
Brewster, NY 10509

ITT Holub Industries
443 Elm St.
Sycamore, IL 60178

Martindale Electric Co.
1375 Hird Av.
Cleveland, OH 44107

Mono-Probe Corp.
4205 Maycrest Av.
Los Angeles, CA 90032

Non-Linear Systems Inc.
533 Stevens Av.
Solana Beach, CA 92075

Pacer Industries Inc.
704 E. Grand Av.
Chippewa Falls, WI 54729

Peschel Instruments Inc.
1412 Viscaya Pkwy
Cape Coral, FL 33904

RFL Industries Inc.
Powerville Rd.
Boonton, NJ 07005

Ross Engineering Corp.
559 Westchester Dr.
Campbell, CA 95008

Rycom Instruments Inc.
9351 E. 59th St.
Raytown, MO 64133

Sangamo Weston Inc.
Schlumberger Division
P.O. Box 3347
Springfield, IL 62714

Sierra Electronic Operation
3885 Bohannon Dr.
Menlo Park, CA 94025

Simpson Electric Co.
Division of American Gage and
  Machine Co.
853 Dundee Av.
Elgin, IL 60120

Snap-On Tools Corp.
2801 80th St.
Kenosha, WI 53140

(A.W.) Sperry Instruments Inc.
245 Marcus Bl.
Hauppauge, NY 11787

Square D Company
P.O. Box 6440
Clearwater, FL 33518

(H.H.) Sticht Co., Inc.
27 Park Pl.
New York, NY 10007

TIF Instruments Inc.
3661 NW 74th St.
Miami, FL 33147

Triplett Corp.
Bluffton, OH 45817

Viz Mfg. Co.
335 E. Price St.
Philadelphia, PA 19144

(The) Von Corp.
P.O. Box 3566G
Birmingham, AL 35211

Western Electro Mechanical
300 Broadway
Oakland, CA 94607

Westinghouse
Relay-Instrument Division
95 Orange St., P.O. Box 606
Newark, NJ 07101

Weston Instruments
614 Frelinghuysen Av.
Newark, NJ 07114

Yokogawa Corp. of America
5 Westchester Plaza
Elmsford, NY 10523

• Volt-Ohm-Ammeters

AEMC Corp.
No. Amer. Dist. Chauvin Arnoux
  Prod.
729 Boylston St.
Boston, MA 02116

Amprobe Instrument
Division of Core Industries Inc.
630 Merrick Rd.
Lynbrook, NY 11563

Associated Research Inc.
8221 N. Kimball Av.
Skokie, IL 60076

(James G.) Biddle Co.
Township Line & Jolly Rds.
Plymouth Meeting, PA 19462

Control Power Systems Inc.
18978 NE 4th Ct.
North Miami Beach, FL 33179

Ecos Electronics Corp.
205 W. Harrison St.
Oak Park, IL 60304

Etcon Corp.
12243 S. 71st Av.
Palos Heights, IL 60463

General Electric Co.
Instrument Products Operation
40 Federal St.
Lynn, MA 01910

General Electric Co.
Instrument Rental Program
1 River Rd.
Schenectady, NY 12345

Hickok Electrical Instruments
10514 Dupont Av.
Cleveland, OH 44108

Hioki New York Corp.
46-16 235th St.
Douglaston, NY 11363

ITT Holub Industries
443 Elm St.
Sycamore, IL 60178

Martindale Electric Co.
1375 Hird Av.
Cleveland, OH 44107

Pacer Industries Inc.
704 E. Grand Av.
Chippewa Falls, WI 54729

RFL Industries Inc.
Powerville Rd.
Boonton, NJ 07005

Ricca-Reddington Instruments, Inc.
1400G NW 65th Av.
Plantation, FL 33313

Ross Engineering Corp.
559 Westchester Dr.
Campbell, CA 95008

Simpson Electric Co.
American Gage and Machine Co.
 Division
853 Dundee Av.
Elgin, IL 60120

Snap-On Tools Corp.
2801 80th St.
Kenosha, WI 53140

(H.H.) Sticht Co., Inc.
27 Park Pl.
New York, NY 10007

(A.W.) Sperry Instruments Inc.
245 Marcus Bl.
Hauppauge, NY 11787

Techni-Tool Inc.
Apollo Rd.
Plymouth Meeting, PA 19462

TIF Instruments Inc.
3661 NW 74th St.
Miami, FL 33147

Triplett Corp.
Bluffton, OH 45817

Viz Mfg. Co.
335 E. Price St.
Philadelphia, PA 19144

Weston Instruments
A Division of Sangamo Weston Inc.
614 Frelinghuysen Av.
Newark, NJ 07114

• **Wattmeters**
AEMC Corp.
No. Amer. Dist. Chauvin Arnoux
 Prod.
729 Boylston St.
Boston, MA 02116

Epic Inc.
150 Nassau St.
New York, NY 10038

General Electric Co.
Instrument Products Operation
40 Federal St.
Lynn, MA 01910

General Electric Co.
Instrument Rental Program
1 River Rd.
Schenectady, NY 12345

Hioki New York Corp.
46-16 235th St.
Douglaston, NY 11363

Knopp Inc.
1307 66th St.
Oakland, CA 94608

Martindale Electric Co.
1375 Hird Av.
Cleveland, OH 44107

RFL Industries Inc.
Powerville Rd.
Boonton, NJ 07005

Sangamo Weston Inc.
Schlumberger Division
P.O. Box 3347
Springfield, IL 62714

Simpson Electric Co.
American Gage and Machine Co.
853 Dundee Av.
Elgin, IL 60120

Square D Company
P.O. Box 6440
Clearwater, FL 33518

Triplett Corp.
Bluffton, OH 45817

Westinghouse
Relay Instrument Division
95 Orange St., P.O. Box 606
Newark, NJ 07101

Weston Instruments
A Division of Sangamo Weston Inc.
614 Frelinghuysen Av.
Newark, NJ 07114

Yokogawa Corp. of America
5 Westchester Plaza
Elmsford, NY 10523

Index